MW00633804

# Materials Horizons: From Nature to Nanomaterials

**Series Editor**

Vijay Kumar Thakur, School of Aerospace, Transport and Manufacturing, Cranfield University, Cranfield, UK

*Materials are* an indispensable part of human civilization since the inception of life on earth. With the passage of time, innumerable new materials have been explored as well as developed and the search for new innovative materials continues briskly. Keeping in mind the immense perspectives of various classes of materials, this series aims at providing a comprehensive collection of works across the breadth of materials research at cutting-edge interface of materials science with physics, chemistry, biology and engineering.

This series covers a galaxy of materials ranging from natural materials to nanomaterials. Some of the topics include but not limited to: biological materials, biomimetic materials, ceramics, composites, coatings, functional materials, glasses, inorganic materials, inorganic-organic hybrids, metals, membranes, magnetic materials, manufacturing of materials, nanomaterials, organic materials and pigments to name a few. The series provides most timely and comprehensive information on advanced synthesis, processing, characterization, manufacturing and applications in a broad range of interdisciplinary fields in science, engineering and technology.

This series accepts both authored and edited works, including textbooks, monographs, reference works, and professional books. The books in this series will provide a deep insight into the state-of-art of *Materials Horizons* and serve students, academic, government and industrial scientists involved in all aspects of materials research.

More information about this series at http://www.springer.com/series/16122

Preetkanwal Singh Bains ·
Sarabjeet Singh Sidhu · Marjan Bahraminasab ·
Chander Prakash
Editors

# Biomaterials in Orthopaedics and Bone Regeneration

## Design and Synthesis

 Springer

*Editors*
Preetkanwal Singh Bains
Department of Mechanical Engineering
Beant College of Engineering and
Technology
Gurdaspur, Punjab, India

Marjan Bahraminasab
Department of Tissue Engineering and
Applied Cell Sciences, School of Medicine
Nervous System Stem Cells Research Center
Semnan University of Medical Sciences
Semnan, Iran

Sarabjeet Singh Sidhu
Department of Mechanical Engineering
Beant College of Engineering and
Technology
Gurdaspur, Punjab, India

Chander Prakash
School of Mechanical Engineering
Lovely Professional University
Jalandhar, Punjab, India

ISSN 2524-5384                ISSN 2524-5392  (electronic)
Materials Horizons: From Nature to Nanomaterials
ISBN 978-981-13-9976-3           ISBN 978-981-13-9977-0  (eBook)
https://doi.org/10.1007/978-981-13-9977-0

This Springer imprint is published by the registered company Springer Nature Singapore Pte Ltd.
The registered company address is: 152 Beach Road, #21-01/04 Gateway East, Singapore 189721, Singapore

# Preface

The aim of this book is to provide information on biomaterials with designed biofunctions, particularly in orthopedics and bone regeneration which involves a large number of injuries and expensive surgeries. These include surface-modified materials, biocomposites, and porous structured biomaterials with similar properties and functions to those of natural bone (i.e., "biomimetic" materials). The research and development being carried out in this particular area has led to novel material designs with improved properties and to distinct cutting-edge routes for the synthesis and processing of such materials. One of the modern fabrication approaches is additive manufacturing (AM) which has been recently used for developing new hybrid biomimetic (tailor-made) materials. The AM techniques can be utilized for making the end-use customized implants for orthopedics and bone scaffolds with a balance in their mechanical and biological properties. This is the focus of Chap. 4 in this book.

The book covers a range of informative chapters which are mainly divided into two applications: orthopedic implants and bone scaffolds. The former such as total joint replacements require characteristics including adequate mechanical strength and high wear resistance. These are usually made of metallic materials (e.g., Ti and its alloys, Co–Cr alloys, and stainless steels) and ceramics (e.g., alumina- and zirconia-based materials). However, to avoid problems such as stress shielding effect and for the tissues to anchor strongly with adjacent bones, these may require designed porosity on their surfaces or specific surface treatments with bioactive materials. Chapters 1, 8, 11, and 14 deal with metallic materials for use as implants and their evaluation through different analyses. Furthermore, Chaps. 3 and 8 discuss on surface treatment of these materials by hydroxyapatite as a bioactive material which is one of the main constituents of bone, and Chap. 10 provides information on the extraction of hydroxyapatite from bovine bone for sustainable development. The latter (i.e., bone scaffolds), however, require different considerations including a porous interconnected structure usually along with biofactors such as stem cells, genes, and proteins to guide the generation of new bone. The porous structure helps in cell nourishment and provides channels for cell migration and surface features for cell attachment. In the meantime, the scaffold structure should provide early mechanical strength to tolerate loads. Chapters 4, 5, 7, and 9 particularly discuss the scaffolds. Chapters 2, 6, and 15 consider the design aspects, selection, optimization, and manufacture of these biomaterials. Furthermore, Chap. 12 explains about the bone and its static and dynamic environments. This information may help in better design of materials for bone substitute.

The book is intended for researchers in the field of orthopedic biomaterials and bone repair and for medical-related practitioners dealing with biomaterial design and manufacture. The book describes the challenges and new opportunities for the potential readers in this research area and can provide a platform for students who seek an interdisciplinary knowledge in the field of newer orthopedic and bone scaffold biomaterials.

Gurdaspur, Punjab, India                         Dr. Preetkanwal Singh Bains
                                                     preetbains84@gmail.com

Gurdaspur, Punjab, India                          Dr. Sarabjeet Singh Sidhu
                                                   sarabjeetsidhu74@gmail.com

Semnan, Iran                                       Dr. Marjan Bahraminasab
                                                  m.bahraminasab@yahoo.com

Jalandhar, Punjab, India                             Dr. Chander Prakash
                                                  chander.mechengg@gmail.com

# Contents

# About the Editors

**Dr. Preetkanwal Singh Bains** is a researcher and faculty in the Department of Mechanical Engineering, Beant College of Engineering and Technology, Gurdaspur, India. His current research interests include Materials Engineering and Advanced Machining Processes. He received his Ph.D. in Mechnical Engineering from IKG Punjab Technical University, India in year 2018. He has contributed numerous innovative and novel manufacturing techniques, for which three patents have been filed in various fields of Mechanical Engineering. He is a member of the Indian Society for Technical Education (ISTE) and American Society of Mechanical Engineers (ASME). He has also contributed many significant publications and served as a reviewer for prominent journals and conferences.

**Dr. Sarabjeet Singh Sidhu** is an Assistant Professor at the Department of Mechanical Engineering, Beant College of Engineering and Technology, Gurdaspur, Punjab, India. He received his Master's of Technology from IKG Punjab Technical University, Jalandhar and Ph.D. from Thapar University, Patiala, India. His research interests include surface modification, and residual stress analysis in metal matrix composites, biomaterials, and non-conventional machining processes. He has published more than 50 technical papers in reputed national and international journals and conferences, and also served as reviewer for various journals. Recently, he was awarded with the Contribution Award by *Journal of Mechanical Science and Technology* (Springer).

**Dr. Marjan Bahraminasab** is an Assistant Professor in the Department of Tissue Engineering and Applied Cell Sciences at Semnan University of Medical sciences. She received her Ph.D. in Biomaterials Engineering from University Putra Malaysia, and is active in the design and manufacture of hybrid biomaterials. Her areas of interest include new biomaterials for hard tissue replacement and repair, orthopedic prostheses, biomaterials design and selection, and biomechanics and finite element analysis.

**Dr. Chander Prakash** is Associate Professor in the School of Mechanical Engineering, Lovely Professional University, Jalandhar, India. He received his Ph.D. in Mechanical Engineering from Panjab University, Chandigarh, India. His area of research include biomaterials and bio-manufacturing. He has more than 11 years of teaching experience and 6 years of research experience. He has contributed extensively in the areas of Titanium and Magnesium based implant with publications appearing in Surface and Coating Technology, Materials and Manufacturing Processes, Journal of Materials Engineering and Performance, Journal of Mechanical Science and Technology, Nanoscience and Nanotechnology Letters, Proceedings of the Institution of Mechanical Engineers, Part B: Journal of Engineering Manufacture. He authored 60 research papers and 10 book chapters. He is also an editor of 2 Books: **Current Trends in Bio-manufacturing** and **Emerging Applications of 3-D Printing in Biomedical Engineering**.

# Chapter 1
# Parametric Evaluation of Medical Grade Titanium Alloy in MWCNTs Mixed Dielectric Using Graphite Electrode

**Preetkanwal Singh Bains, Gurpreet Singh, Amandeep Singh Bhui and Sarabjeet Singh Sidhu**

## 1 Introduction

Electric discharge machining (EDM) in recent decades, has been considered as the most exploited non-conventional machining processes. Herein, electric energy is transformed into heat energy between the tool and the workpiece, utilized for material removal [1, 2]. Widespread applications of EDM could be witnessed in the automobile, aviation, and medical domain. In this thermoelectric technique, material erosion is caused by both the electrodes (tool and workpiece) during the spark generation [3, 4]. A competitive surface topology and accuracy can be achieved by EDM as compared to other conventional machining techniques. The exceptional potential of machining with EDM has been verified with increased strength, higher surface hardness, and formation of intermetallic compounds on the surface [5, 6].

Titanium alloys have been widely used in aerospace as well as biomedical industry owing to their salient physiomechanical and biocompatible attributes. Ti-6Al-4V ($\alpha$-$\beta$) phase alloy is the most preferred metallic Ti alloy in the biomedical applications [7]. However, the presence of vanadium in these alloys besides occupancy of amino acids and proteins in body fluids promotes the corrosive action. This mechanism induces poor osseointegration that leads to cytotoxicity and other allergic reactions

P. S. Bains (✉) · G. Singh · A. S. Bhui · S. S. Sidhu
Department of Mechanical Engineering, Beant College of Engineering and Technology,
Gurdaspur, Punjab, India
e-mail: preetbains84@gmail.com

G. Singh
e-mail: singh.gurpreet191@gmail.com

A. S. Bhui
e-mail: meet_amandeep@yahoo.com

S. S. Sidhu
e-mail: sarabjeetsidhu@yahoo.com

© Springer Nature Singapore Pte Ltd. 2019
P. S. Bains et al. (eds.), *Biomaterials in Orthopaedics and Bone Regeneration*,
Materials Horizons: From Nature to Nanomaterials,
https://doi.org/10.1007/978-981-13-9977-0_1

1

and ultimately failure of the implant. Still, the use of Ti-6Al-4V alloy has been witnessed owing to its admirable human body favorable features. Subsequently, for prolonged superior adhesion between metal, bone and tissue enhancement of surface characteristics of the Ti alloy is of utmost importance [8–10].

Powder mixed electric discharge machining (PMEDM) is one of the recent advancement in material removal processes employed to augment the machining ability and surface characteristics. It has been explored that the use of nano-powder particles in the dielectric enhances the discharge frequency and improves the material removal rate as well as surface quality [11, 12].

Rolling of the graphene sheets into nano-diameter cylinders results in carbon nanotubes (CNTs). CNTs come in two variants, namely single-walled carbon nanotubes (SWCNTs) and Multi-walled carbon nanotubes (MWCNTs). These possess unmatched electrical, mechanical, and thermal properties and thus have a bright future in biomaterial implants [13–15]. CNTs proved to be an excellent substrate material for implants, cell cultivation offering improved cell proliferation and adhesion. CNTs have been found non-toxic at certain concentrations on the titanium surface and inhibit superior cell growth and bone nodule formation [16]. CNTs bestow a well ordered and steady molecular structure, ensuring exceptional features that include higher tensile strength, aspect ratios, and electrical conductivity. In recent times, advancements in synthesis, purification, and functionalization of CNTs have enhanced its virtue as well as its biocompatibility and are currently being actively explored as a material for future bioimplant coatings, composite biomaterials, and biosensors [17].

It has also been concluded by Voge and Stegemann [18] that the adhesion, growth, and differentiation of parent tissue and stem cells can be improved using CNTs. CNTs can also have a significant effect on the evolution of advanced neural implants. Kumar et al. [19] analyzed the MRR and quality of surface of Ti-6Al-4V ELI utilizing EDM process. Current was explored as the most influencing parameter for both MRR and SR. Optimized input variables were 18 A, 40 V, and 100 μs pulse-on duration for achieving highest MRR and least SR. In another experimentation, Das et al. [20] optimized the MRR and SR of EN31 tool steel using EDM process and analyzed that the most influencing parameter for getting highest MRR was the discharge current and later increased steeply with current and pulse-on duration.

Laura et al. [21] put forward the CNTs as appropriate scaffold materials useful for osteoblast proliferation as well as bone formation and witnessed a remarkable alteration in cell morphology in osteoblasts cultured on MWCNTs. It was demonstrated that neutrally charged CNTs sustain osteoblast proliferation and bone-forming functions. The suitability of CNTs for biomaterial applications has been advocated by Harrison and Atala [22] providing a favorable means for tissue engineering along with cell growth. Hirata et al. [23] investigated the formation of bone tissue on MWCNT-coated 3D collagen sponge and concluded that 3D collagen scaffold coated with MWCNT proved to be very effective for bone tissue engineering including cell transplantation. Terada et al. [24] studied the MWCNT-coated layer on the titanium plate for mouse osteoblast-like MC3T3-E1 cells adhesion. The cell adhesion on the MWCNT-coated Ti plate was stronger than on the other plates. Therefore, the

MWCNT coating on the titanium was suggested to be useful for improving cell attachment on titanium implants.

Aleksandra et al. [25] investigated the improvement in the biocompatibility of the commonly used orthopedic implant material, i.e., titanium and its alloys for use in the environment of a living organism, by covering its surface with bioactive layers of MWCNT. Tahsin et al. [26] and Bains et al. [27] in their research work investigated the effect of varying the concentration of SiC powder in powder mixed EDM (PMEDM) on the surface topology, deposition of particles, surface roughness, and structure of top surface formed while machining of light alloys. Maximum deposition of SiC powder was observed for low value (2A) of peak current and higher concentration (20 g/l) of SiC powder in the dielectric. The most significant factors were SiC concentration followed by peak current and pulse-on time.

Mahajan and Sidhu [28] investigated the practicability of various biomaterials in use like Ti alloy, stainless steel, and Cr-Co alloys and the various surface enhancement methods applicable for traditional biomaterials in practice. Lee et al. [29] investigated the ED machined surface of Ti-6Al-4V alloy by varying the duration of the pulse between 10 and 60 $\mu$s. The surface thus produced showed microsurface roughness along with a nano-porous layer of $TiO_2$. An improved bonding, proliferation, and differentiation of MG63 cells were also witnessed. Prakash et al. [30] concluded that EDM possesses great potential in terms of modifying the Ti alloy surface by the formation of oxides and carbides that the new EDMed surface favors biocompatibility, higher surface hardness, improved corrosion resistance, and the development of nano-porous layer. Bhui et al. [31] conducted experimentation by die-sinking EDM using copper electrode on Ti-6Al-4V alloy workpiece with optimized input parameters to obtain high MRR. SEM revealed the improved apatite growth on the surface treated with EDM.

## 2  Materials and Methods

### 2.1  Materials

Grade 5 Ti-6Al-4V alloy in the form of the plate (160 mm × 80 mm × 5 mm) was procured from Baoji Fuyuantong Industry & Trade Co. Ltd., China. For better output results and reduced noise factor, experimentation was performed on two different plates (75 mm × 80 mm × 5 mm) sliced using wire EDM from the main plate. Table 1 tabulates the chemical composition of medical grade Ti-6Al-4V alloy employed in the experimentation.

**Table 1**  Percentage composition of Ti-6Al-4V

| Component | Al | V | O | N | Fe | C | H | Ti |
|---|---|---|---|---|---|---|---|---|
| % | 6.1 | 4.2 | 0.03 | 0.003 | 0.09 | 0.03 | 0.001 | Balance |

The tool electrode material selected for the current experimentation was electrolytic fine/pure graphite. The electrode was machined with the tip diameter of 900 µm. Prior to each experimental run, the electrode face was assured for even flatness and surface finish using emery paper of different grit size, i.e., 350, 600, 1200, and 2000. Multi-walled carbon nanotubes (MWCNTs) procured from United Nanotech, Bangalore, having particle size 10–30 nm and length 10 µ were selected for mixing in the dielectric medium for the powder mixed electric discharge machining (PMEDM) of Ti-6Al-4V. The dielectric used for the experimentation was standard EDM oil.

## 2.2 Pilot Experimentation

Before experimentation, it was necessitated deciding the input parameters for the ED machining of titanium alloy. From the literature review, not much significant work was found regarding the use of graphite as a tool material. For the reason, one-variable-at-a-time (OVAT) technique was employed to observe the machining characteristics of graphite electrode on Ti-6Al-4V. All the trial run as well as experimentation was conducted using reverse polarity of EDM as it is mandatory for the deposition or surface modification of workpiece material. Following Table 2 shows the parametric values of input machining factors taken during the pilot experimentation.

During the trail experimental run, it was seen that with an increase in the current intensity, arcing starts more rapidly leaving a hump on the workpiece and subsequently a hole in tool surface (Fig. 1). It was seen titanium grade 5 cannot be machined with graphite tool using reverse polarity at a higher value of current beyond 4 A. However, pulse-on time also plays a significant role during the

**Table 2** Matrix showing selected values for trial run

| Trial run | Current ($I$) | Pulse-on ($T_{on}$) | Pulse-off ($T_{off}$) | Voltage (V) | Remarks |
|---|---|---|---|---|---|
| 1 | 3 | 90 | 60 | 80 | Arcing |
| 2 | 5 | 90 | 60 | 80 | Arcing |
| 3 | 7 | 90 | 60 | 80 | Arcing |
| 4 | 9 | 90 | 60 | 80 | Arcing |
| 5 | 4 | 90 | 30 | 80 | Arcing |
| 6 | 4 | 90 | 90 | 80 | Arcing |
| 7 | 4 | 90 | 60 | 80 | Arcing |
| 8 | 4 | 90 | 120 | 80 | Arcing |
| 9 | 4 | 45 | 60 | 50 | Arc-free M/C |
| 10 | 4 | 45 | 60 | 80 | Arcing |
| 11 | 4 | 60 | 60 | 60 | Arc-free M/C |

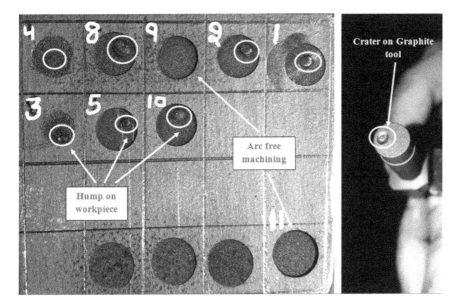

**Fig. 1** Arcing on workpiece and crater on tool

machining because a higher value of $T_{on}$ (beyond 60 μs) also exhibited arcing. Apart from these, the values for $T_{off}$ and gap voltage must be kept large and small, respectively, for the proper and arc-free machining.

The concentration of carbon nanotubes in the dielectric medium was decided by investigating its impact on machining performance. It was varied between 5 and 15 g/l and observed that a higher concentration of MWCNTs in dielectric causes unstable machining. It was due to the reason that MWCNTs contain carbon as its main content causing electrical conductivity and leads to unstable, improper machining. Finally, the concentration was decided as 7 g/l based upon the investigation.

## 2.3 Taguchi $L_{18}$ Design of Experiments

Based upon the observations of pilot experimentation, input machining parameters were chosen for the machining of Ti-6Al-4V with graphite electrode. To reduce the experimental run, Taguchi's methodology of orthogonal arrays was used. Herein, $L_{18}$ array ($2^1$ and $3^4$) was used to design the experiments (Table 4) with the aid of Minitab 17 software. Table 3 shows the control variables, i.e., dielectric medium, current, pulse-on/off duration, and voltage along with their levels selected for the current experimentation.

Aside from these selected parameters, some parameters were kept constant during the experimental work shown in Table 5.

**Table 3** Selected machining variables with levels

| Variables | Units | Levels | | |
|---|---|---|---|---|
| | | Level 1 | Level 2 | Level 3 |
| Dielectric medium | – | EDM oil | EDM oil + MWCNTs | – |
| Current | A | 1 | 2 | 4 |
| Pulse-on duration | μs | 30 | 45 | 60 |
| Pulse-off duration | μs | 60 | 90 | 120 |
| Voltage | V | 30 | 40 | 50 |

**Table 4** $L_{18}$ design matrix for experimentation

| Run | Dielectric medium | Current (A) | $T_{on}$ (μs) | $T_{off}$ (μs) | Voltage (V) |
|---|---|---|---|---|---|
| 1 | EDM oil | 1 | 30 | 60 | 30 |
| 2 | EDM oil | 1 | 45 | 90 | 40 |
| 3 | EDM oil | 1 | 60 | 120 | 50 |
| 4 | EDM oil | 2 | 30 | 60 | 40 |
| 5 | EDM oil | 2 | 45 | 90 | 50 |
| 6 | EDM oil | 2 | 60 | 120 | 30 |
| 7 | EDM oil | 4 | 30 | 90 | 30 |
| 8 | EDM oil | 4 | 45 | 120 | 40 |
| 9 | EDM oil | 4 | 60 | 60 | 50 |
| 10 | EDM oil + MWCNTs | 1 | 30 | 120 | 50 |
| 11 | EDM oil + MWCNTs | 1 | 45 | 60 | 30 |
| 12 | EDM oil + MWCNTs | 1 | 60 | 90 | 40 |
| 13 | EDM oil + MWCNTs | 2 | 30 | 90 | 50 |
| 14 | EDM oil + MWCNTs | 2 | 45 | 120 | 30 |
| 15 | EDM oil + MWCNTs | 2 | 60 | 60 | 40 |
| 16 | EDM oil + MWCNTs | 4 | 30 | 120 | 40 |
| 17 | EDM oil + MWCNTs | 4 | 45 | 60 | 50 |
| 18 | EDM oil + MWCNTs | 4 | 60 | 90 | 30 |

**Table 5** Fixed input controlled parameters

| S. No. | Parameter | Fixed value |
|---|---|---|
| 1 | Powder concentration | 7 g/l |
| 2 | Machining depth | 1 mm |
| 3 | Dielectric used | EDM oil |
| 4 | Polarity | Reverse |

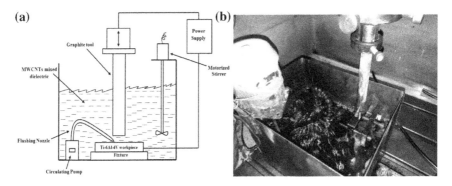

**Fig. 2** **a** Line diagram for the setup; **b** apparatus for PMEDM

## 2.4 Experimentation

Experimentation was conducted on a die-sinker type EDM machine made OSCAR-MAX S645 according to the design matrix shown in Table 4. Out of total 18 experiments, nine were performed in pure dielectric using EDM tank itself, while an in-house developed apparatus consisting stirrer and circulation pump was used to machine the Ti-6Al-4V surface in MWCNTs mixed dielectric medium. Depth of cut (1 mm) was kept constant for all the experiments. Following Fig. 2 illustrates the line diagram and accordingly the in-house setup used for the present work.

## 3 Results and Discussion

MRR, TWR, and surface roughness opted as output parameters for investigating the influence of input machining parameters. Weight of workpiece and tool was measured using digital weighing balance (Citizen made CY220) before and after ED machining for each run to calculate the wear in terms of mg/min. Surface roughness was determined using Mitutoyo SJ400 roughness tester using $R_a$ scale. The design matrix and responses of output parameter are shown in Table 6. Signal-to-noise ratio was used to determine the effects of control variables on output. S/N ratio is defined as signal strength to the error magnitude depending upon the type of response measurement, i.e., "larger is better" or "smaller is better."

## 3.1 ANOVA of S/N Ratios for MRR, TWR, and SR

The output responses were analyzed with the aid of Minitab 17 software employing statistical analysis of variance (ANOVA) of S/N ratios presented in Table 7. Insignificant parameters were pooled during analysis, while most dominating

**Table 6** Experimental design and responses

| Run | Dielectric medium | I (A) | $T_{on}$ (µs) | $T_{off}$ (µs) | (V) | Responses | | | | | | | |
|---|---|---|---|---|---|---|---|---|---|---|---|---|---|
| | | | | | | MRR (mg/min) | | TWR (mg/min) | | SR (µm) | | | |
| | | | | | | Mean | S/N ratio | Mean | S/N ratio | Mean | S/N ratio | | |
| 1 | EDM oil | 1 | 30 | 60 | 30 | 2.00 | 6.0206 | 1.00 | 0.0000 | 0.063 | 24.0132 | | |
| 2 | EDM oil | 1 | 45 | 90 | 40 | 1.75 | 4.8608 | 0.75 | 2.4988 | 0.106 | 19.4939 | | |
| 3 | EDM oil | 1 | 60 | 120 | 50 | 2.00 | 6.0206 | 1.00 | 0.0000 | 0.103 | 19.7433 | | |
| 4 | EDM oil | 2 | 30 | 60 | 40 | 4.84 | 13.6969 | 0.76 | 2.3837 | 0.376 | 8.4962 | | |
| 5 | EDM oil | 2 | 45 | 90 | 50 | 5.03 | 14.0314 | 1.18 | −1.4376 | 0.353 | 9.0445 | | |
| 6 | EDM oil | 2 | 60 | 120 | 30 | 5.03 | 14.0314 | 1.34 | −2.5421 | 0.336 | 9.4732 | | |
| 7 | EDM oil | 4 | 30 | 90 | 30 | 15.90 | 24.0279 | 3.06 | −9.7144 | 1.176 | −1.4081 | | |
| 8 | EDM oil | 4 | 45 | 120 | 40 | 21.09 | 26.4815 | 2.93 | −9.3374 | 0.516 | 5.7470 | | |
| 9 | EDM oil | 4 | 60 | 60 | 50 | 18.58 | 25.3809 | 5.31 | −14.5019 | 1.403 | −2.9412 | | |
| 10 | EDM oil+ MWCNTs | 1 | 30 | 120 | 50 | 2.67 | 8.5302 | 1.00 | 0.0000 | 0.320 | 9.8970 | | |
| 11 | EDM oil+ MWCNTs | 1 | 45 | 60 | 30 | 5.00 | 13.9794 | 0.67 | 3.4785 | 0.093 | 20.6303 | | |
| 12 | EDM oil+ MWCNTs | 1 | 60 | 90 | 40 | 5.67 | 15.0717 | 1.00 | 0.0000 | 0.233 | 12.6529 | | |
| 13 | EDM oil+ MWCNTs | 2 | 30 | 90 | 50 | 7.00 | 16.9020 | 1.33 | −2.4770 | 0.543 | 5.3040 | | |
| 14 | EDM oil+ MWCNTs | 2 | 45 | 120 | 30 | 3.67 | 11.2933 | 1.67 | −4.4543 | 0.196 | 14.1549 | | |
| 15 | EDM oil+ MWCNTs | 2 | 60 | 60 | 40 | 7.67 | 17.6959 | 1.33 | −2.4770 | 0.276 | 11.1818 | | |
| 16 | EDM oil+ MWCNTs | 4 | 30 | 120 | 40 | 6.00 | 15.5630 | 1.67 | −4.4543 | 0.590 | 4.5830 | | |
| 17 | EDM oil+ MWCNTs | 4 | 45 | 60 | 50 | 25.55 | 28.1478 | 3.87 | −11.7542 | 0.123 | 18.2019 | | |
| 18 | EDM oil+ MWCNTs | 4 | 60 | 90 | 30 | 27.78 | 28.8746 | 3.70 | −11.3640 | 0.110 | 19.1721 | | |

**Table 7** Analysis of variance for S/N ratios of outputs

| Factors | DoF | Sum of squares | | | Variance | | | p-value | | | % contribution | | |
|---|---|---|---|---|---|---|---|---|---|---|---|---|---|
| | | MRR | SR | SR | MRR | TWR | SR | MRR | TWR | SR | MRR | TWR | SR |
| Dielectric medium | 1 | 25.695 | # | 32.31 | 25.69 | # | 32.31 | 0.215 | # | 0.441 | 2.44 | # | 3.06 |
| Current | 2 | 757.466 | 405.748 | 364.56 | 378.73 | 202.874 | 182.28 | 0.000* | 0.000* | 0.073* | 72.07 | 77.01 | 34.62 |
| Pulse-on time | 2 | 42.495 | 23.3 | 110.34 | 21.24 | 11.65 | 55.17 | 0.281 | 0.16 | 0.372* | 4.04 | 4.42 | 10.47 |
| Pulse-off time | 2 | 55.986 | # | 27.26 | 27.99 | # | 13.63 | 0.201* | # | 0.765 | 5.32 | # | 2.58 |
| Voltage | 2 | # | 31.025 | 72.01 | # | 15.512 | 36.01 | # | 0.101* | 0.511 | # | 5.88 | 6.83 |
| Error | 8 | 113.594 | 40.023 | 393.78 | 14.19 | 5.003 | 49.22 | | | | | | |
| Total | 17 | 998.35 | 500.547 | 1000.27 | | | | | | | | | |

#Pooled factor; *Significant factor at 95%

process parameters were identified using *p*-value. Discharge current was most significant factor affecting all output responses, i.e., MRR (% contribution: 72.07), TWR (% contribution: 77.01), and SR (% contribution: 34.62). The material removal rate rises with increase in the amount of applied current (4 A) and higher value of pulse-on time (60 μs). Figure 3 illustrates the S/N ratio plot showing effects of input machining parameters on output responses, i.e., MRR, TWR, and SR, respectively.

From Fig. 3, the current was most dominating input factor for all the output variables, whereas the addition of carbon nanotubes in the dielectric medium showed their influence in case of material removal rate and surface roughness.

Based on the current experimentation, the optimum machining conditions for Ti-6Al-4V with graphite tool were 4 A, $T_{on}$ 60 μs, $T_{off}$ 90 μs, 30 V for MRR and 1A, $T_{on}$ 45 μs, $T_{off}$ 60 μs, 30 V for SR, respectively, with the addition of carbon nanotubes in the dielectric medium.

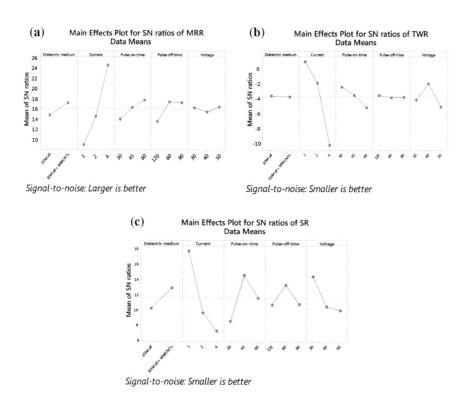

**Fig. 3** Main effects plot for S/N ratio **a** MRR; **b** TWR; **c** surface roughness

## 3.2  Surface Topology and Phase Transformation of ED Machined Samples

To validate the machined surface to be bioactive, machined samples were analyzed for surface morphology and formation of new compounds using scanning microscopy (SEM) and X-ray diffraction (XRD) analysis, respectively. Following Fig. 4 illustrates the microstructure of surface machined by adding carbon nanotubes in the dielectric medium.

Addition of MWCNTs in dielectric not only depicts crack-free surface but also exhibits homogeneous micropores with the deposition of powder particles. These micropores are responsible for proper cell proliferation by providing necessitate adhesion between fractured bone and metallic implant for better biological fixation.

Furthermore, formation of silicides $(V_1.Si_2)$, carbides (TiC), and intermetallic compounds $(Mo_1.Ni_4, Cr_3.O_1, Ni_2.V_1, Al_1.N_1)$ was inspected between theta angles of $10°$–$95°$ via XRD technique. Due to continuous sparking on workpiece submerged in dielectric, compounds were formed by reacting with the hydrocarbons of dielectric fluid and powder mixed.

Figure 5 demonstrates the XRD spectra authenticating the surface modification with the presence of various oxides, carbides, and silicides on machined surface. Formation of such compounds contributes toward bioactivity offering more resistance to wear and corrosion within the individual.

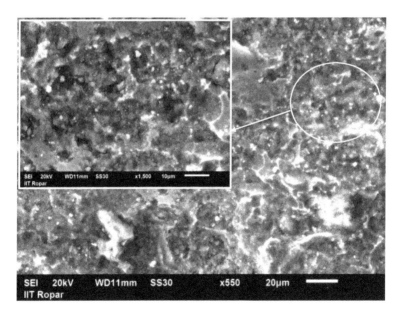

**Fig. 4**  SEM microstructure of powder mixed ED machined surface representing porosity

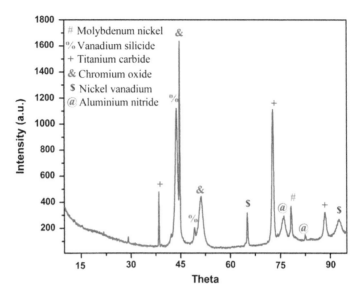

**Fig. 5** XRD spectra of powder mixed ED machined surface showing carbides and intermetallic compounds

# 4 Conclusions

Present work was the parametric investigation for the ED machining of Ti-6Al-4V biomaterial with graphite electrode using reverse polarity. Multi-walled carbon nanotubes were mixed in dielectric medium to examine their influence on machined surface. Following conclusions were made based upon the response observations and surface characteristics of machined samples.

1. Discharge current directly affects the material removal rate. However, current intensity beyond 4 A causes arcing leaving a hump and crater on workpiece and graphite tool, respectively.
2. Addition of carbon nanotubes in dielectric medium exhibits higher value of material removal (27.78 mg/min.) with an incremental of 31.72% compared to material removal rate in pure dielectric.
3. Surface machined at 4 A; $T_{on}$ 30 µs; $T_{off}$ 120 µs; and 40 V showed desired porous structure exhibiting surface roughness of 0.590 µm.
4. Powder concentration in dielectric above 7 g/l causes unstable machining as dielectric becomes conductive due to the colliding of carbon nanotubes and graphite tool.
5. SEM reveals porous surface and deposition of powder particles on machined specimen.
6. X-ray diffraction depicts the surface modification of Ti-6Al-4V with the formation of titanium carbide, chromium oxide, vanadium silicide, and other intermetallic compounds on the machined surface.

# References

1. Bains PS, Sidhu SS, Payal HS (2018) Investigation of magnetic field-assisted EDM of composites. Mater Manuf Processes 33(6):670–675
2. Kumar S, Singh R, Singh TP, Sethi BL (2009) Surface modification by electrical discharge machining: a review. J Mater Process Technol 209(8):3675–3687
3. Bains PS, Sidhu SS, Payal HS (2016) Fabrication and machining of metal matrix composites: a review. Mater Manuf Processes 31(5):553–573
4. Mai C, Hocheng H, Huang S (2012) Advantages of carbon nanotubes in electrical discharge machining. Int J Adv Manuf Technol 59:111–117
5. Bains PS, Singh S, Sidhu SS, Kaur S, Ablyaz TR (2018) Investigation of surface properties of Al–SiC composites in hybrid electrical discharge machining. Futuristic composites. Springer, Berlin, pp 181–196
6. Sidhu SS, Bains PS, Yazdani M, Zolfaniab SH (2018) Application of MCDM techniques on nonconventional machining of composites. Futuristic composites. Springer, Berlin, pp 127–144
7. Peng PW, Ou KL, Lin HC, Pan YN, Wang CH (2010) Effect of electrical-discharging on formation of nanoporous biocompatible layer on titanium. J Alloys Compd 492:625–630
8. Szaraniec B, Pielichowska K, Pac E, Menaszek E (2018) Multifunctional polymer coatings for titanium implants. Mater Sci Eng C 93:950–957
9. Chouirfa H, Bouloussa H, Migonney V, Falentin-Daudre C (2019) Review of titanium surface modification techniques and coatings for antibacterial applications. Acta Biomater 83:37–54
10. Ferraris S, Spriano S (2016) Antibacterial titanium surfaces for medical implants. Mater Sci Eng, C 61:965–978
11. Shabgard M, Khosrozadeh B (2017) Investigation of carbon nanotube added dielectric on the surface characteristics and machining performance of Ti-6Al-4V alloy in EDM process. J Manuf Processes 25:212–219
12. Bains PS, Mahajan R, Sidhu SS, Kaur S (2019) Experimental investigation of abrasive assisted hybrid EDM of Ti-6Al-4V. J Micromanuf. https://doi.org/10.1177/2516598419833498
13. Saito N, Usui Y, Aoki K, Narita N, Shimizu M, Ogiwara N, Nakamura K, Ishigaki N, Kato H, Taruta S, Endo M (2008) Carbon nanotubes for biomaterials in contact with bone. Curr Med Chem 15(5):523–527
14. Raphey VR, Henna TK, Nivitha KP, Mufeedha P, Sabu C, Pramod K (2019) Advanced biomedical applications of carbon nanotube. Mater Sci Eng, C 100:616–630
15. Li XQ, Hou PX, Liu C, Cheng HM (2019) Preparation of metallic single-wall carbon nanotubes. Carbon 147:187–198
16. Mikael PE, Amini AR, Basu J, Arellano-Jimenez MJ, Laurencin CT, Sanders MM, Carter CB, Nukavarapu SP (2014) Functionalized carbon nanotube reinforced scaffolds for bone regenerative engineering: fabrication, in-vitro and in-vivo evaluation. Biomed Mater 9(3):1–13
17. Rivas GA, Rodriguez MC, Rubianes MD, Gutierrez FA (2017) Carbon nanotubes-based electrochemical (bio)sensors for biomarkers. Appl Mater Today 9:566–588
18. Voge CM, Stegemann JP (2011) Carbon nanotubes in neural interfacing applications. J Neutral Eng 8(1):011001
19. Kumar R, Roy S, Gunjan P, Sahoo A, Das RK (2018) Analysis of MRR and surface roughness in machining Ti-6Al-4V ELI titanium alloy using EDM process. Proc Manuf 20:358–364
20. Das MK, Kumar K, Barman TK, Sahoo P (2014) Application of artificial bee colony algorithm for optimization of MRR and surface roughness in EDM of EN31 tool steel. Proc Mater Sci 6:741–751
21. Laura PZ, Bin Z, Hui H, Robert CH (2006) Bone cell proliferation on carbon nanotubes. Nano Lett 6(3):562–567
22. Harrison BS, Atala A (2007) Carbon nanotube applications for tissue engineering: review. Biomaterials 28:344–353
23. Hirata E, Uo M, Takita H, Akasaka T, Watari F, Yokoyama A (2011) Multiwalled carbon nanotubes coating of 3D collagen scaffolds for bone tissue engineering. Carbon 49(10):3284–3291

24. Terada M, Abe S, Akasaka T, Uo M, Kitagawa Y, Watari F (2009) Multiwalled carbon nanotube coating on titanium. Biomed Mater Eng 19:45–52
25. Aleksandra WB, Ewa SZ, Wojciech P, Elzbieta D, Aleksandra B, Marta B (2016) A model of adsorption of albumin on the implant surface titanium and titanium modified carbon coatings (MWCNT-EPD) 2D correlation analysis. J Mol Struct 1–10
26. Tahsin TO, Hamidullah Y, Nihal E, Bülent E (2018) Particle migration and surface modification on Ti6Al4V in SiC powder mixed electrical discharge machining. J Manuf Processes 31:744–758
27. Bains PS, Sidhu SS, Payal HS (2019) Magnetic field influence on surface modifications in powder mixed EDM. Silicon 11(1):415–423
28. Mahajan A, Sidhu SS (2017) Surface modification of metallic biomaterials for enhanced functionality: a review. Mater Technol 33(2):93–105
29. Lee WF, Yang TS, Wu YC, Peng PW (2013) Nanoporous biocompatible layer on Ti-6Al-4V alloys enhanced osteoblast-like cell response. J Exp Clin Med 5(3):92–96
30. Prakash C, Kansal HK, Pabla BS, Puri S, Aggarwal A (2015) Electric discharge machining—a potential choice for surface modification of metallic implants for orthopedic applications: a review. J Eng Manuf 1–23
31. Bhui AS, Singh G, Sidhu SS, Bains PS (2018) Experimental investigation of optimal ED machining parameters for Ti-6Al-4V biomaterial. FU Ser Mech Eng 16(3):337–345

# Chapter 2
# Computational Tailoring of Orthopaedic Biomaterials: Design Principles and Aiding Tools

**Marjan Bahraminasab and Kevin L. Edwards**

## 1 Introduction

The quality of human life can be dramatically improved with the use of biomaterials. Rapidly advancing technologies are allowing new and improved biomaterials to be developed with unprecedented performance behaviour. These materials are being regularly used to make permanent and temporary implants for repairing different parts of the human body, particularly orthopaedic prostheses. The permanently implanted prostheses such as those replacing damaged hip and knee joints are typically made of metals, ceramics and engineering polymers and as the name implies should perform acceptably for a lifetime. The temporarily implanted prostheses such as bone scaffolds are usually based on biodegradable polymers and ceramics that degrade over time. To increase the success of these implants, new biomaterials need to be judiciously designed to rapidly evolve from concept to clinical reality. However, in the design of new biomaterials, costly protocols and long-term experiments are involved (i.e. in vitro and in vivo tests and clinical trials) that must be undertaken before any new product is brought to market (see Fig. 1). This may contain significant commercial risk and uncertainty of outcome for all the efforts made for developing new biomaterials.

To design a new biomaterial, advanced manufacturing techniques [1, 2] and computational approaches [3, 4] should be exploited. The computer modelling of designed materials accelerates novel materials discovery and reduces the risk associated with producing new materials. It does this by providing new opportunities to appreciate how particular structures can be generated in materials, how the

M. Bahraminasab (✉)
Department of Tissue Engineering and Applied Cell Sciences, School of Medicine, Semnan University of Medical Sciences, Semnan, Iran
e-mail: m.bahraminasab@yahoo.com

K. L. Edwards
Institution of Engineering Designers, Courtleigh, Westbury Leigh, Wiltshire BE13 3TA, UK

© Springer Nature Singapore Pte Ltd. 2019
P. S. Bains et al. (eds.), *Biomaterials in Orthopaedics and Bone Regeneration*,
Materials Horizons: From Nature to Nanomaterials,
https://doi.org/10.1007/978-981-13-9977-0_2

**Fig. 1** Schematic of the development process for new biomaterials

structures formed relate to the properties of interest, how the materials respond under real operating conditions and what the design variables of materials are and their optimal values. Also, the consequences of having changes in processing and structure may be assessed and understood, such that the structure–property–performance relationship can be indicated and a basis for specifying the manufacturing approach provided. Meanwhile, the effect of geometry (i.e. shape) variables on the mechanical behaviour of the designed material and the related processing route can be evaluated and it is useful to take into account the concurrent optimization of material and geometry. Further, the computational approaches are capable of estimating the parameters that cannot be readily measured by experiments and of predicting multiple responses, which is the nature of most material design problems, particularly for biomedical applications. Unfortunately, the application of computational methods in the field of biomaterials has been infrequent and fragmented [5], especially when the interactions of biomaterials and the host adjoining tissues are involved. This is probably due to the challenges of establishing suitable computational models that can properly define the complex interactions between biomaterials and living tissues. As a result, research on biomaterial design has mostly been restricted to experimental development, with limited comparisons of several material properties with those of existing materials. Therefore, the generalizing of such trial-and-error approaches on this issue is problematic. However, a more systematic approach is by computational biomaterial design.

## 2 Rules, Procedures and Methods Used in Computational Design of Biomaterials for Orthopaedic Implants

The investigation of biomaterial functionality is a multifaceted problem and depends on many factors and constraints. Therefore, several design rules should be fulfilled, and some tools should be utilized to develop an efficient biomaterial with enhanced properties. One of the main principles is that the structure and characteristics of a biomaterial should be close to those of the replaced or surrounding tissue to avoid any mismatch that may cause failure of the device either prematurely or in the long term. For example, the stiffness of designed biomaterials connecting bone should be similar to that of the bone in order to avoid adverse load transfer and subsequent bone loss. The other important point is to recommend using an inverse approach to develop the desired product as shown in Fig. 2. The conventional approach to new materials design begins with the synthesis of a new material or with a change in

**Fig. 2** Steps involved in
new materials development

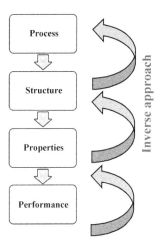

the synthesizing process of an existing material, followed by the characterization
and measurement of the properties. This leads to the question, how far are the characteristics obtained from those required for the specific application? This method,
therefore, is based on a refined trial-and-error scheme in which new materials are
discovered experimentally in laboratories, resulting in spending too much time and
expenditure. The inverse approach, however, starts by identifying the required properties for the desired performance. This then becomes the motivation for indicating
the synthesizing method required to achieve a suitable microstructure that gives the
properties identified. For example, a novel layered material design based on the
inverse approach for replacement of damaged cartilage of knee joint [6] can be made
of three layers: a layer made of wear-resistant metal alloy such as stainless steel at the
articular surface, a layer composed of bioactive ceramic or glass at the bone interface
and an intermediate layer made of a mixture of the two materials placed between the
two layers (Fig. 3).

The other issue in a biomaterials design scenario is that it sometimes contains
multi-scales, such as in tissue engineering for the design of a bone scaffold material. Therefore, the use of computational methods in the field of biomaterials is
rather different but similar design process steps are taken. Most of the computational
biomaterials design conducted to date has been based on finite element analysis

**Fig. 3** Design of layered
material for cartilage
replacement of knee joint

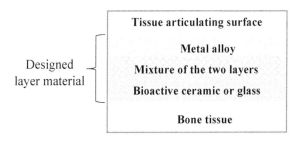

(FEA), which examines new biomaterials by defining calculated properties (as inputs) under different loading conditions.

The following subsections describe the general procedures and techniques used in the computational biomaterials design of hip and knee prostheses because of the numerous and frequent studies conducted and published, and bone scaffolds because of the increasing use in bone surgery.

## 2.1 Computational Methods for Biomaterial Design in Hip and Knee Replacements

In hip and knee joint implants, the main reason for failure can be aseptic loosening of the prosthetic components, predominantly due to wear, micro-motion or stress shielding effect [7, 8]. However, there may be other causes for the implant to not perform successfully or even fail such as peri-prosthetic fracture, mal-positioning, extensor mechanism failure, infection and pain [9–13]. Usually, a complex mechanical and biological process occurs, causing the failure that ends up needing revision surgery, possibly due to implant design [14], patient condition or surgical factors [15, 16]. In implant design, the material is one of the key elements that can contribute to various failure modes of orthopaedic devices, including adverse body reaction to the particulates released from the implant material, failed bonding due to lack of bioactivity, stress shielding phenomenon as a result of high stiffness (modulus of elasticity) and destructive wear caused by a high coefficient of friction. Therefore, material engineers have focused on the development of novel orthopaedic biomaterials [17–20] and some with the help of computational design [21–24].

Almost all new orthopaedic biomaterials have been tailored to possess a hierarchical or functionally graded composition and/or structure in order to achieve several properties simultaneously in a single component that are close to those of the natural organ. A number of mathematical methodologies have been adopted to calculate the gradation in properties and effective material properties (mostly Young's modulus and Poisson's ratio in biomedical applications) of functionally graded materials (FGMs). This includes exponential functions, power law, sigmoid law, volume fraction and rule of mixture, Hashin–Shtrikman-type bounds and Mori–Tanaka-type models [25]. The exponential functions are usually used for an analytical solution. However, the volume fraction and rule of mixture, which provide a more realistic means of representing the continuous FGM properties, may complicate the analytical solution to FGM problems. Therefore, it is usually used in conjunction with finite element modelling (FEM). For examining new orthopaedic biomaterials in real operating situations (i.e. skeletal systems), three-dimensional (3D) models of the bone segment and implant are obtained by computed tomography (CT) data. These models are then imported to finite element (FE) software, the calculated properties of new biomaterial are assigned, and the loading and boundary conditions are applied. To ensure the function of a new biomaterial, several performance metrics and their ideal values are identified based on the target application. These must be obtained from FEA, although it is rather complicated, particularly when simulating the whole bone-implant system and when multiple performance metrics should be gained. The performance metrics that represent aseptic loosening are usually defined to esti-mate wear, implant stability and stress shielding effect (bone loss). For example, the performance metric for the assessment of primary stability of implant component due to lack of osseointegration and bioactivity can be obtained by estimating the micro-motion at the interface of the bone-implant [26–28], where the ideal values would be less than 50 $\mu$m (bone formation) and the non-ideal values are known to exceed 150 $\mu$m (fibrous tissue formation) [29]. The performance metrics can be volumetric/linear wear, maximum stress in contacting surfaces (von Mises and/or contact pressure), wear depth and contact area as measures of wear [30–33], and von Mises stress distribution in the bone (mean, first quartile and standard deviation of stresses), strain energy density and bone mineral density [26, 34–38] as mea-sures of stress shielding in the hip and knee prostheses. When the early design of new biomaterials proves the proposed concept based on the performance metrics (via comparative analyses with the benchmark materials or if applicable by experiments), optimization of new material design variables such as pore size, amount of porosity, volume fraction of constituents and configuration can be done to achieve the best function. The optimization also can be done by FEA via a parametric study.

The use of FEA to investigate the joint system of a new implant design (i.e. biomaterial and implant geometry) is either based on time-independent analyses or time-dependent/time adaptive modelling techniques. In the latter, a parameter of interest (response) is estimated under an initial condition and then used to adapt the FE model by modification of the implant design (geometry and/or material prop-erties) followed by performing a new analysis and recalculating the parameter of interest again. This is an iterative approach that will end when the solution converges or a predefined time period elapses or the bone-implant system fails. These mod-elling techniques contain the following simulations [39], which are also conveniently summarized in Fig. 4.

- Bone remodelling, which is usually used to calculate stress shielding by means of bone mineral density around the femoral component of hip prostheses and tibial and femoral components of knee prostheses [40–43].
- Tissue differentiation that is implemented to envisage the differentiation of granu-lation tissue to fibrous tissue, fibrocartilage or bone based on the micromechanical environment. FEA has been applied to predict tissue differentiation both around implants [44–46], and for modelling the fracture healing process [47, 48].
- Damage accumulation of bone cement, which is conducted according to continuum damage mechanics to estimate fatigue failure of the bone cement [49–51].
- Cement—implant and cement—bone interface de-bonding that is a consecutive detachment process of the cement from the implant based on a static failure crite-rion initially defined [52, 53].
- Wear behaviour is mostly based on Archard's law or a modified version of it. Recently, new wear formulas have been used in computational wear models. In

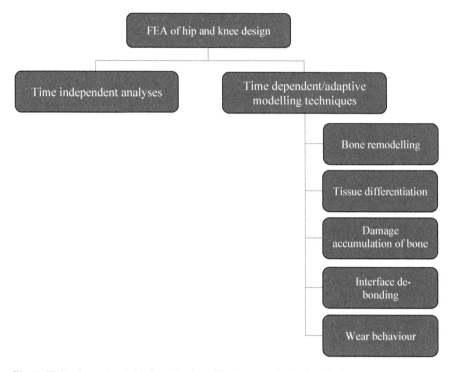

**Fig. 4** Finite element analysis investigation of joint system in implant design

these approaches, wear is related to the contact stress, contact area, wear coefficient and sliding distance. The wear simulation of prosthetic hip and knee joints is aimed at estimating material loss in the softer polyethylene components (i.e. acetabular liner and tibial insert, respectively) and in other types of articular joints such as metal-on-metal joints [30, 54, 55].

## 2.2   Bone Scaffolds

The technology of bone scaffolds is a rapidly developing area of regenerative medicine. A bone scaffold is a temporary porous structural support used to guide the generation of new bone when filling a defect in the skeletal or bone tissue. This is done by transplanting a bio-factor such as stem cells, genes and proteins, within the porous scaffold, which subsequently should be delivered to regenerate natural tissue. It is expected that the bone scaffold tolerates early loads and provides temporary mechanical stability while it degrades over time, so as the defect can be completely filled with new bone tissue. Also, the porous structure is needed for mass transport (means permeability and diffusion) requirements to nourish cells and to provide channels for cell migration and surface features for cell to be attached [56].

In situations where the size of bone defect is large and/or the bone scaffold alone cannot fully support the early loads, an additional temporary implant in the form of metal plate(s) and/or pin(s), or external fixation is used to stabilize the bone. This is later removed when sufficient strength is obtained in the newly formed bone. The usual procedure for using a porous bone scaffold is as follows [57]:

- The mesenchymal stem cells and bone cells are usually taken from the bone marrow and the periosteum of the patient or a suitable donor.
- Since the number of these cells in fresh adult bone marrow is small, culture expansion and cell seeding tasks must be conducted.
- After the bone scaffold implantation, bone growth takes place in the vicinity of the bony defect filled with the bone scaffold.
- Then, scaffold degradation occurs by a mechanism related to bulk erosion arising due to a hydrolysis process.

Several factors including porous architecture (i.e. pore size, pore shape, pore interconnectivity and amount of porosity), resorption kinetics, mechanical and flow properties, biocompatibility, hydrophilicity and pre-seeding influence the success of a designed bone scaffold. One of the challenges in bone scaffold design is achieving a balance between the mechanical function and the bio-factor delivery, which varies between a denser or a more porous bone scaffold [58]. This has encouraged material engineers to develop hierarchical porous bone scaffold (functionally graded porous bone scaffold) structures to achieve the required mechanical properties and mass transport characteristics simultaneously [59]. It has been indicated that heterogeneous bone scaffolds with graded elastic modulus and permeability inversely can minimize the fibrous tissue formation when compared to homogeneous bone scaffolds with constant elastic modulus and permeability [60].

To obtain the hierarchical structure, one can generate libraries of unit cells, as primary components, and assemble them to form the whole desired bone scaffold design as shown in Fig. 5. What is important in the design of bone scaffolds is that the final exterior shape of overall bone scaffold needs to match the complex 3D shape of the bone defect surface, which is achievable by use of medical imaging methods, particularly computed tomography and magnetic resonance imaging (MRI). Structured voxel data sets of patient anatomy, which are based on density distribution, are produced in both CT and MRI. Therefore, the defect shape of interest can be outlined and isolated from these images and a voxel data subset is generated and subsequently used in the design process. The voxel anatomic data can be either converted into solid geometrical models for use in computer-aided design (CAD) [61, 62] or directly used in image-based methods [58, 63, 64]. Next, the internal pore architecture (unit cell) is created and repeated as many times as possible to provide the architecture image. The identified anatomic defect shape is then combined with the internal architecture image through Boolean mathematical techniques, which results in the final bone scaffold design (see Fig. 6). The best functional properties can be obtained by using optimized unit cell architectures that should be as dense as possible for mechanical stability (adequate stiffness or elastic modulus) and as porous as possible for bio-factor delivery. There is consequently conflict in the design requirements because

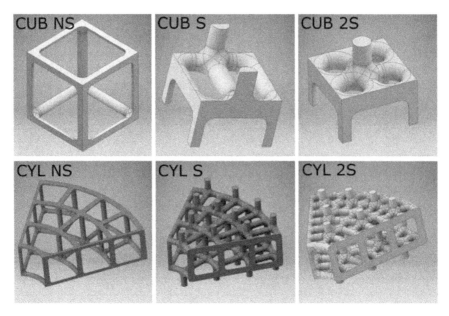

**Fig. 5** Examples of designed unit cells (CUB NS: regular cubic cells, CUB S: single staggered cubic cells, CUB 2S: double staggered cubic cells, CYL NS: regular cylindrical cells, CYL S: single staggered cylindrical cells and CYL 2S: double staggered cylindrical cells) [65]

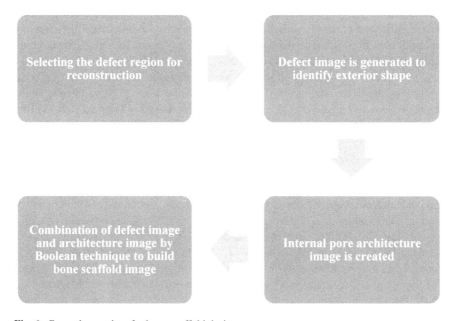

**Fig. 6** General procedure for bone scaffold design

porosity reduces the stiffness. In addition, the porosity used should provide sufficient space for bio-factor seeding and be interconnected for desirable cell nutrition and connective tissue growth. Also, varying the architecture design changes both the elastic modulus and the permeability properties of the bone scaffold. Therefore, bone scaffold design problems involve a number of different design variables and multiple objective functions, thus requiring systematic design tools to be used.

To predict the behaviour of implanted bone scaffolds under various environments, computer modelling and mathematical-based approaches can be very helpful [66]. With respect to the literature, two well-differentiated spatio-temporal scales exist for tissue regeneration of bone scaffolds in vivo, which are the tissue (macroscopic) level and the pore scaffold (microscopic) level. Usually, for the former scale, the macroscopic mechanical properties, and diffusive and flow characteristics are obtained using the theory of homogenization [67], whereas for the latter scale, bone regeneration at the surface of bone scaffolds is simulated by the use of a bone growth model based on bone remodelling theory [68, 69], and bone scaffold degradation can be implemented via a model describing the hydrolysis process [70]. What is important is that when new bone is generated in situ onto the bone scaffold surface or once the bone scaffold is degraded, the macroscopic mechanical properties (e.g. stiffness) also vary. Also, bone growth depends on a particular microscopic mechanical stimulus that can be obtained by solving the macroscopic problem. This indicates that a coupling between the two scales should exist. Therefore, multi-scale modelling is required to define these phenomena properly. Traditionally, modelling of tissue regeneration within bone scaffolds has been confined to one of these scales [60, 70–73]. However, more recent works have applied a micro–macro-mathematical approach that accounts for both scales [74, 75].

There are several tools that can be used in the design process and some of which can be combined with the computational approaches to improve the quality of the output (designed biomaterial).

## 3 Efficient Tools in the Biomaterial Design Process

Computational modelling and analysis play an essential role in understanding the behaviour of new biomaterials. However, it should be born in mind that computational materials models used for goal-oriented design are highly different from those employed traditionally in materials science research activities, which attempt to realize general phenomena [76]. The models for goal-oriented design are aimed at controlling a material system and its optimization to reach one or more desired outcomes to decide between competitive materials (trade-offs) for a specific product. Furthermore, usually, a new biomaterial tailoring starts with general and abstract concept and ends with concrete and detailed configuration. To accomplish this, one can employ several aiding tools such as Design Failure Modes and Effects

Analysis (DFMEA) [77], Quality Function Deployment (QFD) [34, 78], Multi-attribute Decision Making (MADM) [79], Multi-objective Decision Making (MODM) [35, 80] and Design of Experiments (DoE) [81].

The modern material design process usually starts with the identification of design requirements for a specific application, which can be efficiently conducted by DFMEA and QFD. The replacement or modification (tailoring) of a biomaterial might be undertaken if poor performance or failure occurs in the synthesized artificial organ. The failure analysis might indicate that the existing biomaterial could be one of the influential causes. Therefore, it is considered to be a prerequisite to the development of any new material (using the inverse approach). The latter aiding tool (QFD) can help in documenting the voice of the customer (VOC) about the product and translating them into the engineering aspects and technical terms associated with material properties. The customers in the case of biomedical can be patients and complementary team of medical experts and designers. The patients' expectations may include general needs such as lightness of the implant, non-toxicity and strength which possibly can be complemented by the requirements detected by the medical and design team such as bioactivity and osseointegration, appropriate load transfer and low wear between coupling parts. Then, these functional requirements are translated/related to several engineering and technical requirements, i.e. material properties and/or performance metrics. The former aiding tool (DFMEA) can seek for what goes wrong and how the failure of in-use products relates to the material design aspects. For example, what are the causes behind an implant revision? Are the leading causes related to material characteristics such as implant stiffness, hardness, corrosion or bioactivity? Based on this, a set of engineering requirements also can be identified by considering the product needs in terms of the mechanical, physical, chemical, etc., requirements. Thereafter, the material designers need to tailor these properties in the component(s) of interest. Figure 7 shows an example of DFMEA and QFD for femoral component of knee implants.

In biomedical applications, the monolithic materials usually do not satisfy the desired properties and there is a need for manipulating and orienting the constituent materials (multiple materials and/or porosity) in a hierarchical configuration. This involves a selection process for best constituent materials usually with respect to the material properties identified in DFMEA and QFD as selection criteria, which can be accomplished by MADM. For example, a pure metallic component probably suffers from corrosion and mechanically assisted corrosion as well which leads to adverse local tissue reaction. While it is combined with a non-metallic material like ceramic, the concerns of ion release of metallic component can be alleviated. The metal constituent also can decline the brittleness of bulk ceramic component. However, among the very broad list of available metals and ceramics, the choice of appropriate ones is of importance and deals with many criteria including strength, corrosion resistance, fracture toughness, wear resistance, stiffness, biocompatibility issues and many others. Therefore, multi-attribute decision-making approaches can be very useful tools. Meanwhile, this step requires formulating the properties of new material to generate the conceptual material models which can be evaluated by computational approaches such as FEA. When there are a large number of conceptual

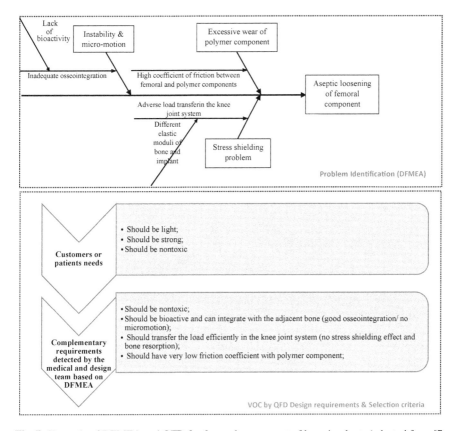

**Fig. 7** Example of DFMEA and QFD for femoral component of knee implants (adapted from [7, 14, 34])

material models, MADM can be used again to select the best material considering the outputs of FEA. There are usually some material design variables (e.g. volume fraction of constituents and direction of gradient), and an optimization procedure is needed to set the material design variables at the appropriate levels. In this stage, FEA can be utilized and effectively combined with DoE to run a parametric study in a reduced number of computational experiments. This enables statistical interpretation on main, quadratic and interaction effects of variables and an adequate search of the design space to save time and reduce cost. Finally, MODM can be applied which is necessary for the biomaterials design problem that seeks to satisfy multiple goals concurrently, e.g. adequate stiffness and degradation properties in bone scaffolds. Thus, optimization may offer more than one solution, which can be tested by simulation and comparison with the benchmark biomaterials. The optimization process has been conducted in the literature, but very little has been done using a multi-objective optimization method systematically by applying DoE and the related software packages.

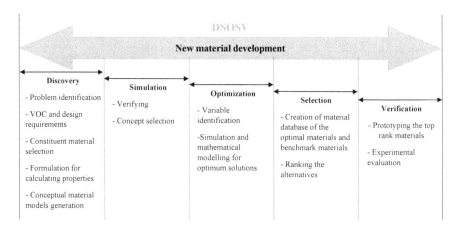

**Fig. 8** Suggested steps of computational biomaterial design

At the end, a selection/ranking process of all alternatives including well-established and optimum computational biomaterials using MADM may require. To ensure the most suitable material, it is essential to provide a material property database with the characteristics of both existing biomaterials and promising ones whose properties are theoretically calculated by computational design. Hence, wide-ranging biomaterials libraries provide a substantial amount of information and enable careful interrogation for a biomaterial with the most preferable properties [82]. There-fore, rather than relying on the intuition and/or experience of biomaterials designers, the most suitable biomaterial can be selected based on rational design criteria. The last stage required is a verifying step which can be done by close corroboration of in vitro and preclinical testing of the devices made of the new biomaterial.

When these tools are used systematically and the above steps are taken in sequence, the design of biomaterial will have "discovery–simulation–optimiza-tion–selection–verification" (DSOSV) steps. Following these steps will enable the designers to do the design process within a structured road map. Figure 8 shows these steps schematically, and Fig. 9 presents the biomaterial design of joint replacements along with the aiding tools.

# 4 Conclusions

Biomaterials design is a multifaceted problem in which several aspects such as the use of inverse material design approach, dependence of required properties to the site of implant in the human body and dealing with macroscopic and microscopic scales should be considered. The biomaterial design process cannot be conducted efficiently if it is left entirely to trial-and-error experimentation. The emphasis was on the role of computational approaches in biomaterials design. Several tools were also introduced to make the biomaterial design process more efficient. The aim of

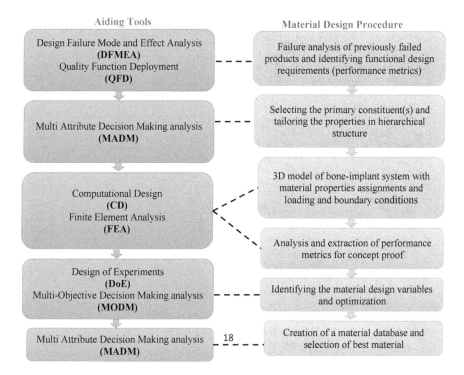

**Fig. 9** Biomaterial design in joint replacements and possible aiding tools

using computational approaches along with these tools is to take into consideration time, cost and quality in the design process. The role of aiding tools is to make the process more logical and intelligible by recognizing the contributions of all inputs, where they fit into the material design process. The motivation for such an approach is to meet a need for biomaterials performance in the context of a given application, where combinations of properties and inherent trade-offs must be simultaneously taken into account.

# References

1. Banerjee S et al (2014) Emerging technologies in arthroplasty: additive manufacturing. J Knee Surg 27(3):185–191
2. Huang Y et al (2015) Additive manufacturing: current state, future potential, gaps and needs, and recommendations. J Manuf Sci Eng 137(1):014001
3. Kanouté P et al (2009) Multiscale methods for composites: a review. Arch Comput Methods Eng 16(1):31–75
4. Boccaccio A et al (2018) A computational approach to the design of scaffolds for bone tissue engineering. Advances in bionanomaterials. Springer, Berlin, pp 111–117
5. Kohn J (2004) New approaches to biomaterials design. Nat Mater 3(11):745–747

6. Nygren M et al (2009) Dual-sided joint implant having a wear resistant surface and a bioactive surface. In: International application published under the patent cooperation treaty (PCT), W.I.P. Organization, Editor. Google Patents, US
7. Bahraminasab M et al (2012) Aseptic loosening of femoral components—a review of current and future trends in materials used. Mater Des 42:459–470
8. Melvin JS et al (2014) Early failures in total hip arthroplasty—a changing paradigm. J Arthroplasty 29(6):1285–1288
9. Brown NM et al (2015) Extensor mechanism allograft reconstruction for extensor mechanism failure following total knee arthroplasty. J Bone Joint Surg 97(4):279–283
10. Morrey MC (2015) Revision total knee arthroplasty: management of periprosthetic femur fracture around total knee arthroplasty. Complex primary and revision total knee arthroplasty. Springer, Berlin, pp 129–142
11. Nishii T et al (2015) Fluctuation of cup orientation during press-fit insertion: a possible cause of malpositioning. J Arthroplasty 30(10):1847–1851
12. Liddle AD, Rodríguez-Merchán EC (2015) Periprosthetic fractures. Total knee arthroplasty. Springer, Berlin, pp 219–227
13. Jiranek WA et al (2015) Surgical treatment of prosthetic joint infections of the hip and knee: changing paradigms? J Arthroplasty 30(6):912–918
14. Bahraminasab M et al (2013) Aseptic loosening of femoral components-materials engineering and design considerations. Mater Des 44:155–163
15. Katz JN et al (2012) Twelve-year risk of revision after primary total hip replacement in the US Medicare population. J Bone Joint Surg 94(20):1825–1832
16. Marius N et al (2015) Biomaterials view on the complications associated with hip resurfacing arthroplasty. In: Advanced materials research, vol 1114
17. Oshkour A et al (2015) Mechanical and physical behaviour of newly developed functionally graded materials and composites of stainless steel 316L with calcium silicate and hydroxyapatite. J Mech Behav Biomed Mater 49:321–331
18. Doni Z et al (2015) Tribocorrosion behaviour of hot pressed CoCrMo—HAP biocomposites. Tribol Int 91:221–227
19. Dehaghani MT, Ahmadian M, Beni BH (2015) Fabrication and characterization of porous Co–Cr–Mo/58S bioglass nano-composite by using NH 4 HCO 3 as space-holder. Mater Des 88:406–413
20. Patel AK, Balani K (2015) Dispersion fraction enhances cellular growth of carbon nanotube and aluminum oxide reinforced ultrahigh molecular weight polyethylene biocomposites. Mat Sci Eng C 46:504–513
21. Bahraminasab M et al (2013) Material tailoring of the femoral component in a total knee replacement to reduce the problem of aseptic loosening. Mater Des 52:441–451
22. Enab TA, Bondok NE (2013) Material selection in the design of the tibia tray component of cemented artificial knee using finite element method. Mater Des 44:454–460
23. Hedia H et al (2014) A new design of cemented stem using functionally graded materials (FGM). Bio-med Mater Eng 24(3):1575–1588
24. Mehboob H, Chang S-H (2015) Optimal design of a functionally graded biodegradable composite bone plate by using the Taguchi method and finite element analysis. Compos Struct 119:166–173
25. Gupta A, Talha M (2015) Recent development in modeling and analysis of functionally graded materials and structures. Progr Aerosp Sci 79:1–14
26. Oshkour A et al (2015) Parametric study of radial functionally graded femoral prostheses with different geometries. Meccanica 1–22
27. Oshkour AA et al (2013) Finite element analysis on longitudinal and radial functionally graded femoral prosthesis. Int J Numer Methods Biomed Eng 29(12):1412–1427
28. Oshkour AA et al (2015) Effect of geometrical parameters on the performance of longitudinal functionally graded femoral prostheses. Artif Organs 39(2):156–164
29. Taylor M, Barrett DS, Deffenbaugh D (2012) Influence of loading and activity on the primary stability of cementless tibial trays. J Orthop Res 30(9):1362–1368

30. Willing R, Kim IY (2009) Three dimensional shape optimization of total knee replacements for reduced wear. Struct Multi Optim 38(4):405–414
31. Abdelgaied A et al (2011) Computational wear prediction of artificial knee joints based on a new wear law and formulation. J Biomech 44(6):1108–1116
32. Mattei L, Di Puccio F, Ciulli E (2013) A comparative study of wear laws for soft-on-hard hip implants using a mathematical wear model. Tribol Int 63:66–77
33. Netter J et al (2015) Prediction of wear in crosslinked polyethylene unicompartmental knee arthroplasty. Lubricants 3(2):381–393
34. Jahan A, Bahraminasab M (2015) Multicriteria decision analysis in improving quality of design in femoral component of knee prostheses: influence of interface geometry and material. In: Advances in materials science and engineering
35. Bahraminasab M et al (2014) Multi-objective design optimization of functionally graded material for the femoral component of a total knee replacement. Mater Des 53:159–173
36. Bahraminasab M et al (2014) On the influence of shape and material used for the femoral component pegs in knee prostheses for reducing the problem of aseptic loosening. Mater Des 55:416–428
37. Van Lenthe GH et al (2002) Stemmed femoral knee prostheses: effects of prosthetic design and fixation on bone loss. Acta Orthop 73(6):630–637
38. Completo A et al (2009) Relationship of design features of stemmed tibial knee prosthesis with stress shielding and end-of-stem pain. Mater Des 30(4):1391–1397
39. Taylor M, Prendergast PJ (2015) Four decades of finite element analysis of orthopaedic devices: where are we now and what are the opportunities? J Biomech 48(5):767–778
40. Rezaei F et al (2015) Carbon/PEEK composite materials as an alternative for stainless steel/titanium hip prosthesis: a finite element study. Australas Phys Eng Sci Med 1–12
41. Gillies RM et al (2007) Adaptive bone remodelling of all polyethylene unicompartmental tibial bearings. ANZ J Surg 77(1–2):69–72
42. Andersen MR, Petersen MM (2015) Adaptive bone remodeling of the femoral bone after tumor resection arthroplasty with an uncemented proximally hydroxyapatite-coated stem. J Clin Densitometry 19(2):202–207
43. Pérez M et al (2014) Bone remodeling in the resurfaced femoral head: Effect of cement mantle thickness and interface characteristics. Med Eng Phys 36(2):185–195
44. Mukherjee K, Gupta S (2015) Bone ingrowth around porous-coated acetabular implant: a three-dimensional finite element study using mechanoregulatory algorithm. Biomech Model Mechanobiol 1–15
45. Waide V et al (2004) Modelling the fibrous tissue layer in cemented hip replacements: experimental and finite element methods. J Biomech 37(1):13–26
46. Puthumanapully PK, Browne M (2011) Tissue differentiation around a short stemmed metaphyseal loading implant employing a modified mechanoregulatory algorithm: a finite element study. J Orthop Res 29(5):787–794
47. Miramini S et al (2015) Computational simulation of the early stage of bone healing under different configurations of locking compression plates. Comput Methods Biomech Biomed Eng 18(8):900–913
48. Miramini S et al (2015) The relationship between interfragmentary movement and cell differentiation in early fracture healing under locking plate fixation. Australas Phys Eng Sci Med 1–11
49. Stolk J et al (2002) Finite element and experimental models of cemented hip joint reconstructions can produce similar bone and cement strains in pre-clinical tests. J Biomech 35(4):499–510
50. Coultrup OJ et al (2010) Computational assessment of the effect of polyethylene wear rate, mantle thickness, and porosity on the mechanical failure of the acetabular cement mantle. J Orthop Res 28(5):565–570
51. Bouziane M et al (2015) Analysis of the behaviour of cracks emanating from bone inclusion and ordinary cracks in the cement mantle of total hip prosthesis. J Braz Soc Mech Sci Eng 37(1):11–19

52. Caouette C et al (2015) Influence of the stem fixation scenario on load transfer in a hip resurfacing arthroplasty with a biomimetic stem. J Mech Behav Biomed Mater 45:90–100
53. Van de Groes S, de Waal-Malefijt M, Verdonschot N (2014) Probability of mechanical loosening of the femoral component in high flexion total knee arthroplasty can be reduced by rather simple surgical techniques. Knee 21(1):209–215
54. Abdelgaied A et al (2014) The effect of insert conformity and material on total knee replacement wear. Proc Inst Mech Eng Part H J Eng Med 228(1):98–106
55. Gao L, Dowson D, Hewson RW (2015) Predictive wear modeling of the articulating metal-on-metal hip replacements. J Biomed Mater Res Part B Appl Biomater 105(3):497–506
56. Bellucci D et al (2011) A new generation of scaffolds for bone tissue engineering. In: Advances in science and technology. Trans Tech Publ
57. Sanz-Herrera J, García-Aznar J, Doblaré M (2009) On scaffold designing for bone regeneration: a computational multiscale approach. Acta Biomater 5(1):219–229
58. Hollister SJ et al (2005) Engineering craniofacial scaffolds. Orthod Craniofac Res 8(3):162–173
59. Hollister SJ (2005) Porous scaffold design for tissue engineering. Nat Mater 4(7):518–524
60. Kelly DJ, Prendergast PJ (2006) Prediction of the optimal mechanical properties for a scaffold used in osteochondral defect repair. Tissue Eng 12(9):2509–2519
61. Hutmacher DW, Sittinger M, Risbud MV (2004) Scaffold-based tissue engineering: rationale for computer-aided design and solid free-form fabrication systems. Trends Biotechnol 22(7):354–362
62. Chu T-MG et al (2002) Mechanical and in vivo performance of hydroxyapatite implants with controlled architectures. Biomaterials 23(5):1283–1293
63. Hollister SJ et al (2000) An image-based approach for designing and manufacturing craniofacial scaffolds. Int J Oral Maxillofac Surg 29(1):67–71
64. Feinberg SE et al (2001) Image-based biomimetic approach to reconstruction of the temporomandibular joint. Cells Tissues Organs 169(3):309–321
65. Dallago M et al (2018) Fatigue and biological properties of Ti-6Al-4V ELI cellular structures with variously arranged cubic cells made by selective laser melting. J Mech Behav Biomed Mater 78:381–394
66. Sengers BG et al (2007) Computational modelling of cell spreading and tissue regeneration in porous scaffolds. Biomaterials 28(10):1926–1940
67. Sanchez-Palencia E, Zaoui A (1987) Homogenization techniques for composite media. In: Homogenization techniques for composite media
68. Beaupré G, Orr T, Carter D (1990) An approach for time-dependent bone modeling and remodeling—theoretical development. J Orthop Res 8(5):651–661
69. Van Lenthe G, De Waal Malefijt M, Huiskes R (1997) Stress shielding after total knee replacement may cause bone resorption in the distal femur. J Bone Joint Surg Br 79(1):117–122
70. Adachi T et al (2006) Framework for optimal design of porous scaffold microstructure by computational simulation of bone regeneration. Biomaterials 27(21):3964–3972
71. Sanz-Herrera J, Garcia-Aznar J, Doblare M (2008) A mathematical model for bone tissue regeneration inside a specific type of scaffold. Biomech Model Mechanobiol 7(5):355–366
72. Hollister SJ, Maddox R, Taboas JM (2002) Optimal design and fabrication of scaffolds to mimic tissue properties and satisfy biological constraints. Biomaterials 23(20):4095–4103
73. Adachi T et al (2001) Trabecular surface remodeling simulation for cancellous bone using microstructural voxel finite element models. J Biomech Eng 123(5):403–409
74. Sanz-Herrera J, García-Aznar J, Doblaré M (2008) Micro–macro numerical modelling of bone regeneration in tissue engineering. Comput Methods Appl Mech Eng 197(33):3092–3107
75. Chen Y, Zhou S, Li Q (2011) Microstructure design of biodegradable scaffold and its effect on tissue regeneration. Biomaterials 32(22):5003–5014
76. Kuehmann C, Olson G (2009) Computational materials design and engineering. Mater Sci Technol 25(4):472–478
77. Thapa N, Prayson M, Goswami T (2015) Case studies in engineering failure analysis
78. Santiago A et al (2015) Design of an impulsion prosthetic system for prosthetic foot. In: VI Latin American congress on biomedical engineering CLAIB 2014, Paraná, Argentina 29, 30 and 31 Oct 2014. Springer

79. Jahan A, Edwards KL, Bahraminasab M (2016) Multi-criteria decision analysis for supporting the selection of engineering materials in product design. Butterworth-Heinemann, Boston
80. Alaimo G et al (2017) Multi-objective optimization of nitinol stent design. Med Eng Phys 47:13–24
81. Aherwar A, Singh A, Patnaik A (2016) Study on mechanical and wear characterization of novel Co30Cr4Mo biomedical alloy with added nickel under dry and wet sliding conditions using Taguchi approach. Proc Inst Mech Eng Part L J Mater Des Appl. https://doi.org/10.1177/1464420716638112
82. Curtarolo S et al (2013) The high-throughput highway to computational materials design. Nat Mater 12(3):191–201

# Chapter 3
# EDM Surface Treatment: An Enhanced Biocompatible Interface

**Amit Mahajan, Sarabjeet Singh Sidhu and Timur Ablyaz**

## 1 Introduction

In earlier times, biomaterials were utilized in therapeutic devices but since the last decade, their level of sophistication has increased considerably resulting in their extensive usage in the biomedical field. In the starting of the twentieth century, natural materials such as wood which was used in mend or built up of the body, inaugurate to be replaced by fabricated polymers, ceramics, and metal amalgams [1]. The pivotal properties requisite for materials utilized in orthopedic and orthodontic purpose are to be of high corrosion and wear resistance and should demonstrate excellent biocompatibility and osseointegration to avert conflicting biological reaction [2]. Pre-eminent metallic alloys employing biomaterials as implant materials comprise Co alloys, Ti alloys, and stainless steels [3]. These materials offered improved performance, augmented functionality, and better reproducibility than their naturally acquired analog [4]. The cobalt–chromium-(Co–Cr)-based alloy F75 (Co–Cr–Mo) has been extensively utilized in total joint replacements and proved their applications in the orthopedic field [5]. These alloys have tremendous mechanical properties, adequate corrosion resistance, excellent wear resistance, and competent biocompatibility [6]. It was observed from some cases of cobalt alloys that the ions of Co, Cr, and Mo were incorporate with body tissues through suspension of the passive oxide layer and

A. Mahajan
IKGPTU, Kapurthala 144603, Punjab, India
e-mail: amitmahajan291@gmail.com

S. S. Sidhu (✉)
Department of Mechanical Engineering, Beant College of Engineering and Technology, Gurdaspur 143521, India
e-mail: sarabjeetsidhu@yahoo.com

T. Ablyaz
Perm National Research Polytechnic University, Perm, Russia
e-mail: lowrider11-13-11@mail.ru

© Springer Nature Singapore Pte Ltd. 2019
P. S. Bains et al. (eds.), *Biomaterials in Orthopaedics and Bone Regeneration*,
Materials Horizons: From Nature to Nanomaterials,
https://doi.org/10.1007/978-981-13-9977-0_3

corrosion wear process [7, 8]. In order to modify the surfaces and mechanical properties, nonconventional machining methods were utilized because of their excellent machining results with high accuracy [9, 10]. EDM was discovered for machining of complex materials and alloys among all the nonconventional methods [11, 12]. Mahajan et al. [13, 14] presented the study on bioimplant's surface treatment by EDM. They reported in their further research that EDM is the promising technique to construct the orthopedic implants. Bigerelle et al. [15] and Peng et al. [16] nominated EDM, as an appropriate surface treatment for biological material applications owing to improved morphology, promising cell proliferation, favorable biocompatibility, and the propitious tissue implant relation. The EDM-treated samples are also hemocompatible in nature claimed by Mastud et al. [17]. They considered vibration-assisted reverse micro-electrical discharge machining (R-MEDM) process for surface textured of Ti-6Al-4V. Their study revealed that the R-MEDM-treated samples show texture surfaces that had favorable results with blood interaction.

However, to the best of our knowledge, machining of a Co–Cr alloy by EDM to improve their biocompatibility is hardly reported and there is no study available which consider HeLa cell lines on Co–Cr surfaces for investigating the cytocompatibility of the samples. This study is focused on the impact of EDM technique on the biological behavior of the Co–Cr alloy implant.

## 2   Materials and Methods

In this study, the mechanical grade Co–Cr–Mo alloy (ASTM-F75) (Supplier: DePuy Synthes, Massachusetts, USA) was utilized for experimentation. This alloy has enormous applications in orthopedic field and the chemical composition of the procured alloy is cobalt with Chromium: 28.5%, Molybdenum: 6%, Silicon: 0.7%, Manganese: 0.5%, Nickel: 0.25%, Carbon: 0.22%, Iron: 0.2%, Titanium: 0.01% and Phosphorus: 0.02%. The pictographic diagram and microstructure of the femoral part of a knee implant are presented in Fig. 1.

The finely grained graphite electrode (particles sized 5 μm) was utilized for surface treatment of the specimen in die sinking EDM (make OSCARMAX, Taiwan) with negative polarity [the workpiece (−) and tool electrode (+)]. The abundance of electric sparks is generated in a sequence of time through EDM which is an electrothermal machining process [18]. Through electric sparks, a large amount of heat is generated which removes the material from specimens resulting in easier to machine the hard and tough surfaces [19, 20]. In order to achieve the appropriate stability in the EDM process, the dielectric medium (i.e., EDM oil) was utilized. The current at 16A, pulse on time at 200 μs with constant off time (20 μs), and voltage (140 V) were considered as the final set of experiment. As beyond this, unsteady arcing was observed on the specimen that damages the surface. The EDM process parameters, their corresponding levels, and schematic diagram of EDM setup are illustrated in Table 1 and Fig. 2.

**Fig. 1** Pictorial view and microstructure of the femoral component of knee joint

**Table 1** Process parameters and their values

| Working parameters | Description |
|---|---|
| Electrode, Diameter (mm) | Graphite, 10 |
| Workpiece | Cr–Co–Mo alloy |
| Polarity | Electrode (+), workpiece (−) |
| Voltage (V) | 140 |
| Current (A) | 5,10,16 |
| Pulse on (μs) | 60,150,200 |
| Pulse off time (μs) | 20 |
| Machining time (min) | 30 |
| Dielectric | EDM oil |
| Dielectric flushing pressure | 0.06 kg/cm$^2$ |

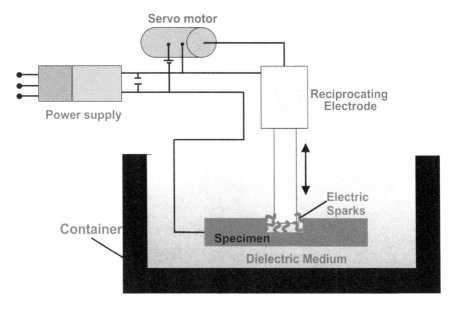

**Fig. 2** Schematic arrangement of EDM setup

**Table 2** L-9 experimental design and % cell viability response

| Trial no. | $I$ (A) | $P_{on}$ ($\mu$s) | % Cell viability |
|---|---|---|---|
| 1 | 5 | 60 | 46 |
| 2 | 5 | 150 | 39 |
| 3 | 5 | 200 | 80 |
| 4 | 10 | 60 | 51 |
| 5 | 10 | 150 | 67 |
| 6 | 10 | 200 | 28 |
| 7 | 16 | 60 | 76 |
| 8 | 16 | 150 | 33 |
| 9 | 16 | 200 | 40 |

Three stages of current and pulse on time were selected and Taguchi's L-9 experimental array was intended to examine the cytotoxicity responses of the specimens as indicated in Table 2.

## 2.1 In Vitro Cytocompatibility Study

The In-vitro cytocompatibility of EDMed Co-Cr alloy was analyzed using MTT [3-(4,5-Dimethylthiazol-2-yl)-2, 5-Diphenyltetrazolium Bromide] [$C_{18}H_{16}BrN_5S$] assay in HeLa cell lines (Cervical cancer cells) by the following reported protocol.

The untreated sample was considered as a positive control and Dulbecco's Modified Eagle Medium (DMEM) (Dealer: Himedia, Mumbai, India) on glass coverslip was considered as a negative control due to its cell maintenance capability and observed with 100% cell viability. At first, EDM samples were placed in an Autoclave (Accumax, New Delhi, India) for half an hour at 121 °C where bacteria, algae, fungi, etc., were removed from samples by moist heat sterilization. Further, HeLa cells were seeded in Minimum Essential Medium Eagle (MEM), accompany with Earle's salts, L-glutamine, nonessential amino acids, sodium bicarbonate, sodium pyruvate, 10% fetal bovine serum (FBS), 100 U ml$^{-1}$ penicillin, and 100 mg ml$^{-1}$ streptomycin (PAA Laboratories GmbH, Austria). Further, these HeLa cells were preserved at 37 °C under the atmospheric condition of 5% $CO_2$. After that, the cell harvesting process was conducted by employing a 0.25% w/v trypsin—Ethylenediaminetetraacetic acid (EDTA) solution. Further, the metal samples (size-10 mm dia, 2.5 mm thickness) were placed in a 24-well plate (from Corning, USA) where 10,000 cells were brought in contact with these samples. After the 48-h incubation period, the plate was taken out from the incubator and removed metal samples from the plate and further put these samples in new 24 well culture plates [21]. The amount of the cell viability is analyzed by the transformation of MTT to purple-colored formazan due to metabolically active cells. Further, these outcomes were compared with negative control and positive control samples. Later on, by utilizing dimethyl sulfoxide (DMSO), the cells were solubilized, where spectrophotometric technique at 570 nm was used to measure the absorbance of solubilized formazan reagent [22]. Three trials were conducted for each experiment. The cell viability responses of EDMed samples were illustrated in Table 2. The %cell viability was calculated by:

$$\% \text{ cell viability} \frac{AB_s}{AB_c} * 100,$$

where $AB_s$ is the absorbance of the testing samples and $AB_c$ is the absorbance of the control.

## 3 Results and Discussion

Experimentation was conducted according to L-9 experimental array with two independent variables such as current and pulse on time. The experimental layout and results are given in Table 2. The in vitro cytotoxicity of electrical discharge treated and untreated samples were calculated by evaluating the % cellular viability as illustrated in Fig. 3. The HeLa cell viability on Co–Cr substrates was observed to be significantly less in comparison with the negative control sample (100% cell viability). Also, the insignificant difference of cell viability responses was observed between the EDMed samples 6, 8 and positive control (unmachined sample—26% cell viability). While, the EDMed samples 1, 2, 4, 5 and 9 verified higher cell viability as compared to the untreated sample and were observed within the span of 39–67%.

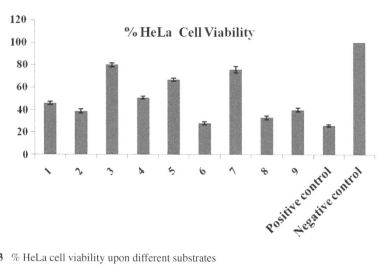

**Fig. 3**  % HeLa cell viability upon different substrates

Although, sample 3 (~80%) and sample 7 (~76%) were observed with significantly higher cell viability as compared to positive control. For sample 3, the cell viability was approximately 2.45 times more than the unmachined surface. Figure 4 represents the cell toxicity response of sample 3 under the microscope. The maximum number

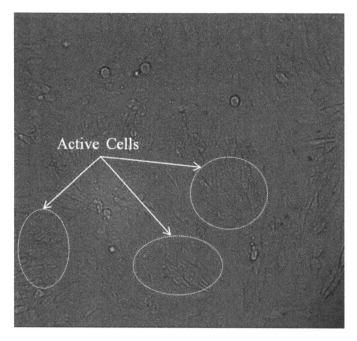

**Fig. 4**  Microscope image represents the cytotoxicity response upon sample 3

of living cells was observed on the sample surface that confirms the biocompatibility of the EDMed sample. The results depicted improved the cytocompatibility response of EDM-treated samples as compared to an untreated sample that admires the EDMed implants in the biomedical field.

## 4 Conclusions

In this study, the Co–Cr alloy surface was altered by EDM to examine the biological behavior. The in-vitro biocompatibility responses of the samples were scrutinized by cytocompatibility test. The outcomes of the study demonstrated the higher HeLa cell viability on EDMed substrates as compared to the untreated specimen. It was also observed that the substrate surface modified by EDM at 5 A current and 200 µs pulse on time (Trial 3) enhance the biocompatibility of the specimen. Thus, EDM-treated Co–Cr implants can be indulged in the biomedical field.

**Acknowledgements** The authors thank IKG Punjab Technical University, Kapurthala, for its support to this research work.
**Ethical approval** No animal testing was performed during this study.

## References

1. Huebsch N, Mooney DJ (2009) Inspiration and application in the evolution of biomaterials. Nature 462(7272):426–432
2. Long M, Rack HJ (1998) Titanium alloys in total joint replacement—a materials science perspective. Biomaterials 19:1621–1639
3. Walia B, Sarao TPS, Grewal JS (2015) Corrosion studies of thermal Sprayed HA and HA/Al$_2$O$_3$-TiO$_2$ bond coating on 316L stainless steel. Trends Biomater Artif Organs 29(3):245–252
4. Ristic M, Manic M, Misic D, Kosanovic M, Mitkovic M (2017) Implant material selection using expert system. FU Mech Eng 15(1):133–144
5. Hyslop DJS, Abdelkader AM, Cox A, Fray DJ (2010) Electrochemical synthesis of a biomedically important Co–Cr alloy. Acta Mater 58(8):3124–3130
6. Tarzia V, Bottio T, Testolin L, Gerosa G (2007) Extended (31 years) durability of Starr-Edwards prosthesis in mitral position. Interact Cardiovasc Thoracic Surg 6(4):570–571
7. Posada OM, Tate RJ, Grant MH (2015) Effects of CoCr metal wear debris generated from metal-on-metal hip implants and Co ions on human monocyte-like U937 cells. Toxicol In Vitro 29(2):271–280
8. Okazaki Y, Emiko G (2005) Comparison of metal release from various metallic biomaterials in vitro. Biomaterials 26(1):11–21
9. Bains PS, Sidhu SS, Payal HS, Kaur S (2019) Magnetic field influence on surface modifications in powder mixed EDM. Silicon 11(1):415–423
10. Klocke F, Zeis M, Klink A, Veselovac D (2013) Technological and economical comparison of roughing strategies via milling, EDM and ECM for titanium-and nickelbased blisks. Proc CIRP 2(1):98–101

11. Kumar S, Singh R, Batish A, Singh TP (2012) Electric discharge machining of titanium and its alloys: a review. Int J Mach Mach Mater 11(1):84–111
12. Bains PS, Sidhu SS, Payal HS (2018) Investigation of magnetic field-assisted EDM of composites. Mater Manfact Process 33(6):670–675
13. Mahajan A, Sidhu SS (2017) Surface modification of metallic biomaterials for enhanced functionality: a review. Mater Technol 33(2):1–13
14. Mahajan A, Sidhu SS (2018) Enhancing biocompatibility of Co–Cr alloy implants via electrical discharge process. Mater Technol 33(8):1–8
15. Bigerelle M, Anselme K, Noel B, Ruderman I, Hardouin P, loat A (2002) Improvement in the morphology of Ti-basedsurfaces: a new process to increase in vitro human osteoblast response. Biomaterials 23(7):1563–1577
16. Peng PW, Ou KL, Lin HC, Pan YN, Wang CH (2010) Effect of electrical-discharging on formation of nanoporousbiocompatible layer on titanium. J Alloys Compd 492:625–630
17. Mastud S, Garg M, Singh RK, Samuel J, Joshi H (2012) Experimental characterization of vibration-assisted reverse micro electrical discharge machining (EDM) for surface texturing. In: Proceeding of ASME international manufacturing science and engineering conference, pp 439–448
18. Bains PS, Sidhu SS, Payal HS (2016) Study of magnetic field-assisted ED machining of metal matrix composites. Mater Manfu Processes 31(14):1889–1894
19. Bhui AS, Singh G, Sidhu SS, Bains PS (2018) Experimental investigation of optimal ED machining parameters for Ti-6Al-4 V biomaterial. FU Mech Eng 16(3):337–345
20. Bains PS, Mahajan R, Sidhu SS, Kaur S (2019) Experimental investigation of abrasive assisted hybrid EDM of Ti-6Al-4V. J Micromanuf. https://doi.org/10.1177/2516598419833498
21. Longster GH, Buckley T, Sikorski J, Derrick-Tovey LA (1972) Scanning electron microscope studies of red cell morphology. Vox Sang 22(2):161–170
22. Ahmadian S, Barar J, Saei AA, Fakhree MA, Omidi Y (2009) Cellular toxicity of nanogenomedicine in MCF-7 cell line: MTT assay. J Vis Exp 3(26):e1191. https://doi.org/10.3791/1191

# Chapter 4
# Development of Cellular Construction for the Jaw Bone Defects Replacement by Selective Laser Melting

**Polina Kilina, Lyudmila Sirotenko, Evgeniy Morozov, Timur Ablyaz and Karim Muratov**

## 1 Introduction

Reconstructive surgery is connected with the introduction of medical constructions into the tissues with the aim to reconstruct the lost or defective pieces of bones. In this case, the material of the implant should have the biological compatibility, low inflexibility, high strength, adhesive surface, and be not the cause of immunological rejection [1, 2]. For the effective invasion of the bony tissue to the implant cells, the geometrics of microstructure of the implant material should be characterized by the optimal size of pores [3].

Technology of selective laser melting (SLM) has the greatest potential in solving the problem of bone defect regeneration. SLM technologies based on 3D modeling are currently introduced in the most science-intensive areas of science and production including different fields of medicine, particularly implantology [4, 5]. This method allows producing implant to a high degree of accuracy. Such implant presents the form similar to that of the replaced defect and has a cellular microstructure which promotes the intensification of biological processes and reduction in time for the bone tissue regeneration [6].

P. Kilina · L. Sirotenko · E. Morozov · T. Ablyaz (✉) · K. Muratov
Perm National Research Polytechnic University, Perm, Russia
e-mail: lowrider11-13-11@mail.ru

P. Kilina
e-mail: sunshine0420@mail.ru

L. Sirotenko
e-mail: sirotenko@pstu.ru

E. Morozov
e-mail: morozov.laser@gmail.com

K. Muratov
e-mail: karimur_80@mail.ru

© Springer Nature Singapore Pte Ltd. 2019
P. S. Bains et al. (eds.), *Biomaterials in Orthopaedics and Bone Regeneration*,
Materials Horizons: From Nature to Nanomaterials,
https://doi.org/10.1007/978-981-13-9977-0_4

Properties of the cellular structures depend both on the basic material, geometrical structure, and the technique of their production. Comparing to conventional methods of powder metallurgy, selective laser melting technique is preferable for making products with complex spatial geometry and uniform regulated structure [7].

At present, the application of Ti6Al4V alloys having the required biomechanical characteristics, biocompatibility, and corrosion resistance in biomedicine is paid much attention. Titanium alloys save their properties in the process of interaction with the aggressive biological environment of an organism; they do not have toxic influence on tissues and completely meet the demands made to the implants [8, 9].

The basis of the cellular material lattice frame is the bridges on the properties and structure of which depend on Physical–mechanical properties of the material in whole. Such material is assigned for the implant production by SLM technique. Experimental study of melting of isolated tracks simulating bridges geometry of the constructed high-porous cellular implant is the basic research for determining efficient laser fusion regimes which provide required geometrics and structural characteristics of the track (the lack of incomplete penetration of particle in the central zone and in periphery, uniformity of melting, correspondence of the simulated, and experimental track dimensions).

The objective of the work is the optimization of geometric structure and process parameters of making implants with the required elastic and strength properties on the basis of titanium alloys by the application of selective laser melting technology for the replacement of bone structures' defects.

To achieve this aim, it has been solved the following tasks:

- Development of 3D models of samples with the cellular macrostructure similar to the trabecular (spongy) bone organization.
- Selection of efficient process parameters of laser melting aimed at providing necessary geometrics and macrostructural parameters of cellular construction.
- Experimental study of elastic and strength properties of highly porous Ti6Al4V alloy as well as verification of its Physical–mechanical properties correspondence to those of bone tissue.
- Production of experimental cellular samples of implants from Ti6Al4V alloy by laser melting technique with the aim of maxillofacial bone defects replacement.
- Research of the processes of bone structures' regeneration after the introduction of implants to the laboratory animals.

## 2   Materials, Equipment, and Results of Research

### 2.1   Modeling of Cellular Constructions

Three-dimensional modeling of the experimental sample lattice frame geometry and of implants taking place before the process of selective laser melting was made in

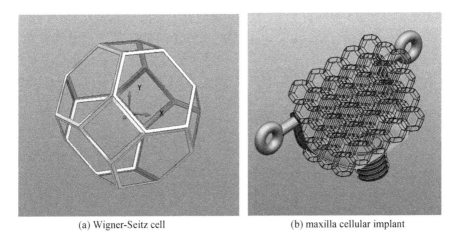

| (a) Wigner-Seitz cell | (b) maxilla cellular implant |

**Fig. 1**  Three-dimensional models of cellular structures

the PowerShape (Delcam) system. Further processing of the model STL files was conducted in the Magics (Materialize) program: Positioning of 3D model on the substrate was carried out, and it was constructed capabilities between platform and part for its further unfailing removal and for the support of the task geometry in the process of construction; step of dissection into layers was equal to 30 μm.

Wigner–Seitz cells of 1, 2, and 3 mm and diameter of the bridge equal to 0.2; 0.25; and 0.3 mm were the base of the construction. Moreover, the cells were similar to the structural elements of trabecular bone (Fig. 1a). All wireframe objects represent hard leaf bodies; cells fill maximum space in the limits of the given volume and have a minimal surface area. In the construction of the maxilla implant, it was provided the elements of support pieces (Fig. 1b).

## 2.2   Ti6Al4V Powder Materials Characteristic

Examination of powder and obtained samples was carried out by the Altami MET 5D reflected light microscope. Ti6Al4V medium-dispersed metal powder of primarily spherical form with the average size of particles equal to 30 μm was used as the basic material (Fig. 2). Such characteristics of powder as bulk density, tap density, and specific surface depend on the form and size of particles, dispersity, and degree of surface maturity. Bulk density and tap density amounted to 2.499 and 2.746 g/cm$^3$ correspondingly. Specific surface was determined by means of Sorbi 4.1 unit and gaseous adsorption technique. Nitrogen was used as gas adsorbate at a temperature of—196 °C (temperature of liquid nitrogen), and BET constant amounted to 24.48; the value of specific surface—0.14 m$^2$/g. Energy dispersive X-ray fluorescence analysis of the powder sample was carried out by EDX-800HS Shimadzu spectrometer

**Fig. 2** Ti6Al4V metal powder

**Table 1** Elemental composition of the basic powder Ti6Al4V

|          | Percentage of metal elements (%) | | | | | |
|----------|------|------|------|------|------|----------|
|          | Al   | V    | Si   | Cu   | Fe   | Ti       |
| Analysis | 6.29 | 4.09 | –    | –    | –    | The rest |

in a cuvette with polypropylene window, the environment of analysis is air, and the collimator is 5 mm. Elemental composition of the basic powder is given in Table 1.

Results of the research show that powder is characterized by a high degree of purity; it is free from impurities which the important demand made by the medical industry.

## 2.3 Influence of Laser Melting Conditions on Geometrics of Bridges in Cellular Materials

Experimental samples were made by the method of selective laser melting on the Realizer SLM 50 unit equipped with the high-power laser (100 W). Production of the sample was carried out in the argon medium; scanning was made with the 90° change of direction from layer to layer. The input parameters for the unit are: power $P$, specified time $t$, and distance between points $L$.

To optimize the structure of cellular materials, several experimental studies were performed with the aim to determine the influence of laser melting conditions on geometrics of the bridges in cellular materials. It was carried out a two-factor experiment according to the method of orthogonal central composite planning. Power values varied in the range of $P$ from 25 to 100 W, specified time t from 20 to 60 μs, and distance between points $L$ from 5 to 25 μm. Single tracks were obtained; their surface structure and uniformity of melting were estimated. Press fitting for further making of microsections was made by LECO PR4X priming press. The image of single tracks with $P = 37.5$ is presented in Fig. 3. Specified time and distance between points were varied.

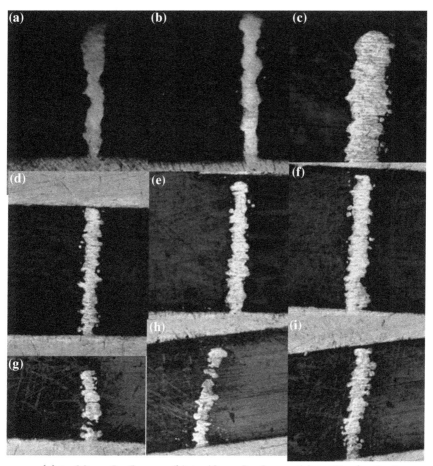

(a) t =20 μs; L=5 μm ;  (b) t =40 μs; L=5 μm;  (c) t=60 mks; L=5 μm;
(d) t=20 μs; L=15 μm;  (e) t=40 μs; L=15 μm;  (f) t=60 μs; L=15 μm;
(g) t=20 μs; L=25 μm;  (h)t=40 μs; L=25 μm;(i) t=60 μs; L=25 μm

**Fig. 3**  Lateral microsection of the track at $P = 37.5$ W

Equations of the track width ($h$) dependence on conditions of melting are of the form:

$$h(P, t) = 0.31 \cdot P - 0.70 \cdot t + 0.06 \cdot P \cdot t + 77.81 \tag{1}$$

$$h(P, L) = 4.73 \cdot P - 1.48 \cdot L + 0.18 \cdot P \cdot L + 136.56 \tag{2}$$

$$h(t, L) = 5.13 \cdot t + 1.5 \cdot L - 0.22 \cdot t \cdot L + 55.66 \tag{3}$$

where $P$—power of laser emission, $t$—specific time, $L$—distance between points, and $h$—width of the track.

Dependences of the track width from the power of laser emission, specific time, and distance between tracks are given in Fig. 4. In the diagram, the area of acceptable values of width is colored with yellow. When the values are less than 150 μm, the width values difference has a great effect, thinning of bridges takes place, and this fact is prohibitive. Values higher than 350 μm lead to formation of low and closed porosity in the cells. For instance, when the width of the bridge is 400 μm and the diameter of the cell is 1 mm, porosity of the sample accounts only to 18%.

It was chosen the mode panel supporting the acceptable values of the track width: emission power $P = 37.5$–50 W, specific time $t = 20$ μs, distance between tracks $L = 5$ μm and power of emission $P = 37.5$–75 W, specific time $t = 40$–60 μs, and

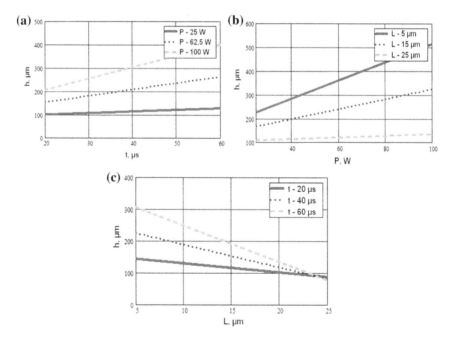

**Fig. 4** Diagrams of the width of the track $h$ depending on the conditions of laser melting

distance between tracks $L = 15$ µm. These parameters of processing provide the uniformity of melting, the lack of poorly melted particles in the central zone of the track and in the periphery, correspondence of the simulated and derived geometry features.

Further research was carried out in the regime of the following values: emission power $P = 37.5$ W, specific time $t = 40$ µs, and distance between tracks $L = 15$ µm.

## 2.4   Ti6Al4V Cellular Materials Compression Test

In the process of variation of cell's diameters D from 1 to 3 mm and diameters of bridge t from 0.2 to 0.3 mm, cellular samples with different porosity from 0.5 to 0.97 were obtained. Universal floor standing electromechanical testing machine Instron 5885H was used for the mechanical test; the number of repeats of every test was $k = 3$. The diagram of compressive stress–strain state of cellular samples with the cells diameter equal to 2 mm and diameter of bridges equal to 0.3 mm is given in Fig. 5; the diagrams for the rest combination of diameters and widths of bridges have a similar nature. At the initial stage, deformation becomes higher along with the increase of loading; then, the stress approaches the value when brittle failure of some bridges with further compression (interlocking) of the structure takes place; every peak corresponds to the failure of the bridges separate blocks.

Compression stress and modulus of elasticity dependency on the diameters of cell and bridge are presented in Figs. 6 and 7, and also, it is shown in the frames the range of mechanical properties for trabecular bone in compression. Average values of the human trabecular bone ultimate strength are in the limits of 10–200 MPa, and the value of modulus of elasticity is in the range of 0.02–2 GPa (green field in the diagram). For the ultimate strength of the maxilla trabecular bone, average values in

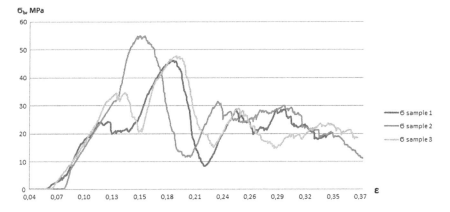

**Fig. 5** Diagram of compression of the samples with the diameter of cells equal to 2 mm and diameter of bridges equal to 0.3 mm

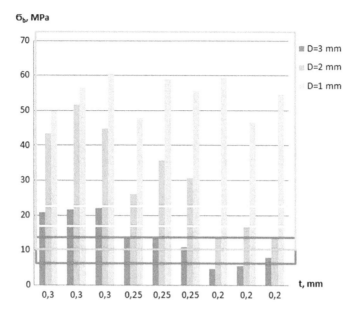

**Fig. 6** Dependence of ultimate strength of $\sigma_b$ construction on the diameter of the cell $D$ and the bridge $t$

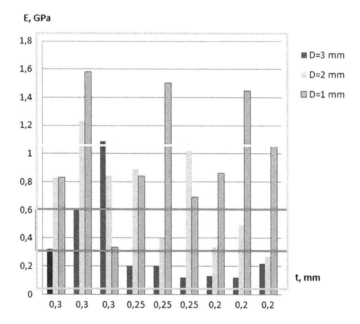

**Fig. 7** Dependence of modulus of strength of $E$ construction on the diameter of the cell $D$ and the bridge $t$

**Fig. 8** Dependence of
ultimate compression
strength $\sigma_b$ on the diameter
of the cell $D$ and the bridge $t$

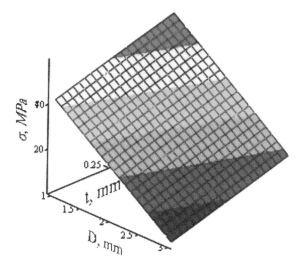

the lengthwise direction equal to 16–22 MPa and for the modulus of elasticity—from 0.6 to 1.05 GPa—are the most characteristic ones (yellow field of the diagram). Ultimate strength in the crosswise direction assumes the values in the range of 7–13 MPa, modulus of strength—from 0.375 to 0.6 GPa (red field of the diagram) [10, 11].

Ultimate strength of the obtained samples assumes the value in the range of 4.6–60.6 MPa, while the modulus of elasticity is changed from 0.12 to 1.58 GPa. The values of maximum stress in construction should exceed ultimate strength of the bone, while the modulus of elasticity should correspond or be lower than the modulus of elasticity of the bone. Stress condition is fulfilled for the cells with a diameter of 2 mm and bridges of 0.25 and 0.3 mm, as well for the cells with a diameter of 1 mm. Modulus of elasticity condition is satisfied for the cells of 2 and 3 mm.

The next relationship was derived in the process of regression analysis:

$$\sigma_b = 29.5 - 20.29 \cdot D + 165 \cdot t \tag{4}$$

where $D$—diameter of cell, $t$—diameter of bridge, and $\sigma_b$—ultimate strength. According to the results of this analysis, it could be made a conclusion that compression strength of the cellular structure becomes higher when the size of the unit cell decreases, while the diameter of the bridge becomes larger (Fig. 8).

## 2.5  Testing of Cellular Structures Implantation into Laboratory Animals

The next stage was the fabrication of maxilla implants with the aim to study the processes of regeneration of bone structures after their implantation in laboratory

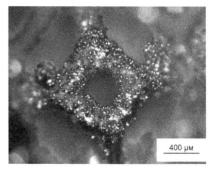

(a) Cellular implant construction;              (b) Surface structure of the cell

**Fig. 9**  Maxilla cellular implant

animals (Fig. 9). All experiments were carried out in accordance with the Order of the Ministry of Public Health of the USSR from 12.08.1977, No. 755 and "European convention on vertebrate animals' protection used for experiments and in other scientific aims" from 18.03.1986. The text has been changed according to the protocol regulations (ETS No. 170), after its coming into force on December 2, 2005. Cellular structures of approximately similar porosity with the diameter of cell equal to 3 mm and diameter of the bridge equal to 0.3 mm, diameter of cell—2 mm and diameter of bridge—0.25 mm; diameter of cell—1 mm and diameter of bridge—0.2 mm were fabricated. The sample had a developed surface which is preferable for cytoadherence (Fig. 9b). Developed cellular structures were implanted in the laboratory animals' maxillae in the Central Research Laboratory of Acad. E. A. Vagner Perm State Medical University of RF Ministry of Public Health (Fig. 10).

Local and total reaction to the implant was evident only at the initial stage (8–10 days). In two weeks, the first group of animals was taken out of the experiment. Six mounts were analyzed in the light microscope with 50 times enlargement. It was

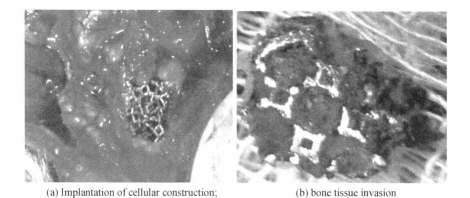

(a) Implantation of cellular construction;              (b) bone tissue invasion

**Fig. 10**  Implantation of maxilla cellular structure

stated that the processes of osteoblasts' formation began in all cells. In three months, there were prepared 30 samples from the removed out of the cells tissue with aim to study reparative features. All samples were divided into three groups depending on the size of cells. During the microscopic research, it was revealed that tissue closely surrounds on the outside all three types of implants. In the implant with the cells of 1 mm, there is a loss of tissue in the central cells; cells of the 2 mm are filled with implant in full scale and the implant with the cells of 3 mm fills the cellular structure for 2/3 (Fig. 11).

Estimation of the local reaction on the operative intervention was made during six months. Also it was estimated general condition of the animals according to the integral indices (behavior, appetite, temperature reaction). Local complications in the process of operative wound healing as well as implants rejection were not observed; general reaction was normal. In six month, all cells were filled with bone tissue with different degree of fixation. Thus, cellular structure promoted quicker invasion of bone tissue in the product of implantation.

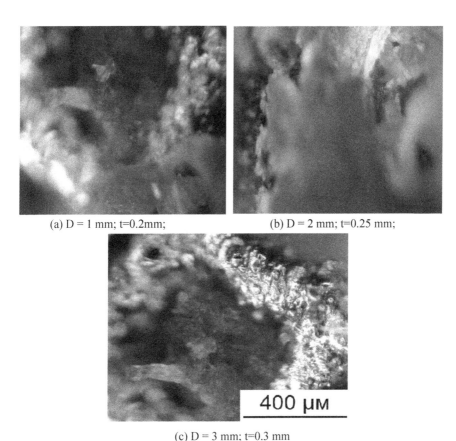

(a) D = 1 mm; t=0.2mm;          (b) D = 2 mm; t=0.25 mm;

(c) D = 3 mm; t=0.3 mm

**Fig. 11** Formation of tissue in the cells of implants

# 3 Conclusions

On the basis of 3D modeling, construction of different porosity implants from Wigner-Seitz cells replicating the structure of trabecular bone tissue has been developed. It has been determined the conditions of laser melting providing the uniformity of bridge melting and geometrics required for shaping cells of open porosity (power of laser emission—37.5 W, specific time—40 μs, distance between point—15 μm). Calculation of ultimate strength and modulus of elasticity in cellular structure materials has been made. Using SLM technique, there has been produced test models of cellular samples fabricated from Ti6Al4V powder with geometrics of microstructure similar to that of bone tissue. Experimental study of bone structures regeneration after the introduction of implants in bone tissue of laboratory animals has been carried out. The time of engraftment made up 3–6 months. It has been determined that construction of implant with the diameter of bridges equal to 250 μm, diameter of cells equal to 2 mm and porosity of 90% imitating the structure of maxilla bone tissue is characterized by compression strength $\sigma_b = 35$ MPa and modulus of elasticity $E = 0.8$ GPa, corresponds to the properties of bone tissue, provides its accelerated invasion in the cells of implant, and promotes the reduction in regeneration period.

**Acknowledgements** The authors officially thank V. P. Vasilyuk, candidate of medical science, associate professor of the department of dental and oral surgery of Acad. E. A. Vagner PSMU, for the experiments on implantation of cellular structures into laboratory animals. Research has been carried out under financial support of the Ministry of Education and Science of RF according to the state-guaranteed order 11.9716.2017/8.9.
**Ethical approval** All applicable international, national, and/or institutional guidelines for the care and use of animals were followed.

# References

1. Hollander DA, Walter MV, Wirtz T, Sellei R, Schmidt-Rohlfing B, Paar O, Erli HJ (2006) Structural, mechanical and in vitro characterization of individually structured Ti-6Al-4V produced by direct laser forming. Biomaterials 27:955–963
2. Parthasarathy J, Starly B, Raman S, Christensen A (2010) Mechanical evaluation of porous titanium (Ti6Al4V) structures with electron beam melting (EBM). J Mech Behav Biomed Mater 3(3):249–259
3. Wang X, Xu S, Zhou S, Xu W, Leary M, Choong P, Qian M, Brandt M, Xie YM (2016) Topological design and additive manufacturing of porous metals for bone scaffolds and orthopaedic implants: a review. Biomaterials 83:127–141
4. Murr LE, Gaytan SM, Medina F, Lopez H, Martinez E, Machado I, Hernandez H, Martinez L, Lopez MI, Wicker RB, Bracke J (2010) Next-generation biomedical implants using additive manufacturing of complex, cellular and functional mesh arrays. Philos Trans 368:1999–2032
5. Mullen L, Stamp RC, Brooks WK, Jones E, Sutcliffe CJ (2009) Selective laser melting: a unit cell approach for the manufacture of porous, titanium, bone in-growth constructs, suitable for orthopedic applications. J Biomed Mater Res B Appl Biomater 89(2):325–334

6. Pattanayak DK, Fukuda A, Matsushita T, Takemoto M, Fujibayashi S, Sasaki K, Nishida N, Nakamura T, Kokubo T (2011) Bioactive Ti metal analogous to human cancellous bone: fabrication by selective laser melting and chemical treatments. Acta Biomater 7(3):1398–1406
7. Vasilyuk VP, Straub GI, Kachergus SA, Kosarev PV, Asanovich MO (2014) Experimental justification of using innovation of technologies in manufacturing implants with cellular structure for replacing bone defects in facial skeleton (preliminary results). Family Health—21st century 2:42–54
8. Bhui AS, Singh G, Sidhu SS, Bains PS (2018) Experimental investigation of optimal ED machining parameters for Ti-6Al-4V biomaterial. FU Mech Eng 16(3):337–345
9. Bains PS, Mahajan R, Sidhu SS, Kaur S (2019) Experimental investigation of abrasive assisted hybrid EDM of Ti-6Al-4V. J Micromanuf. https://doi.org/10.1177/2516598419833498
10. Vasilyuk VP, Shtraube G, Chetvertnykh VA (2017) Optimization of surgery trearme of the anti-total defects of the jaws with the use of honeycomb structures. Mod Probl Sci Educ 3:1–8
11. Olson SA, Marsh JL, Anderson DD, Latta Pe LL (2012) Designing a biomechanics investigation: choosing the right model. J Orthop Trauma 26(12):672–677

# Chapter 5
# Squeeze Film Bearing Characteristics for Synovial Joint Applications

T. V. V. L. N. Rao, Ahmad Majdi Abdul Rani and Geetha Manivasagam

## Nomenclature

| | |
|---|---|
| $B$ | Width of parallel plate, m |
| $C$ | Partial journal bearing radial clearance, m |
| $h, H$ | Film thickness, m; $H = h/h_o$ for parallel plate; $H = h/C$ for partial journal bearing |
| $h_o$ | Reference film thickness, m |
| $k_1, K_1$ | Permeability of porous layer in region $I$, m$^2$; $K_1 = k_1/h_0^2$ for parallel plate; $K_1 = k_1/C^2$ for partial journal bearing |
| $L$ | Length of parallel plate; length of partial journal bearing, m |
| $p$ | Pressure distribution, N/m$^2$; $P = ph_0^3/\mu L^2(-\mathrm{d}h/\mathrm{d}t)$ for parallel plate, $P = pC^2/\mu R^2(\mathrm{d}\varepsilon/\mathrm{d}t)$ for partial journal bearing |
| $q, Q$ | Flow rate (volume) per unit length, m$^2$/s; $Q = q/L(-\mathrm{d}h/\mathrm{d}t)$ for parallel plate; $Q = q/RC(\mathrm{d}\varepsilon/\mathrm{d}t)$ for partial journal bearing |
| $R$ | Journal radius, m |
| $u_i, U_i$ | Fluid velocity in surface (or porous) layer region I, couple stress fluid film region II, surface layer region III, respectively, m/s; $U_i =$ |

T. V. V. L. N. Rao (✉)
Department of Mechanical Engineering, SRM Institute of Science and Technology,
Kattankulathur 603203, India
e-mail: tvvlnrao@gmail.com

A. M. A. Rani
Department of Mechanical Engineering, Universiti Teknologi PETRONAS, 32610 Seri Iskandar,
Malaysia
e-mail: majdi@utp.edu.my

G. Manivasagam
Centre for Biomaterials, Cellular and Molecular Theranostics (CBCMT),
Vellore Institute of Technology, Vellore, India
e-mail: geethamanivasagam@vit.ac.in

© Springer Nature Singapore Pte Ltd. 2019
P. S. Bains et al. (eds.), *Biomaterials in Orthopaedics and Bone Regeneration*,
Materials Horizons: From Nature to Nanomaterials,
https://doi.org/10.1007/978-981-13-9977-0_5

$u_i/[(L/h_o)(-\mathrm{d}h/\mathrm{d}t)]$ along $x$ direction in parallel plate; $U_i =$ $u_i/[R(\mathrm{d}\varepsilon/\mathrm{d}t)]$ along $\theta$ direction in partial journal bearing, $i = 1, 2, 3$

$u_j, U_j$    Fluid velocity at the interface of surface (or porous) layer region I and couple stress fluid film region II, couple stress fluid film region II and surface layer region III, respectively, m/s; $U_j = u_j/[(L/h_o)(-\mathrm{d}h/\mathrm{d}t)]$ along $x$ direction in parallel plate; $U_j = u_j/[R(\mathrm{d}\varepsilon/\mathrm{d}t)]$ along $\theta$ direction in partial journal bearing, $j = 12, 23$

$w, W$    Static load, N; $W = wh_0^3/\mu L^3 B(-\mathrm{d}h/\mathrm{d}t)$ for parallel plate; $W = wC^2/\mu R^3 L(\mathrm{d}\varepsilon/\mathrm{d}t)$ for partial journal bearing

$x, X$    $X$ direction coordinate, m; $X = x/L$ for parallel plate; $\theta = x/R$ for partial journal bearing

$y, Y$    $Y$ direction coordinate, m; $Y = y/h_o$ for parallel plate; $Y = y/C$ for partial journal bearing

## Greek Letters

$\delta_i, \Delta_i$    Thickness of surface (or porous) layer region I, couple stress fluid film region II, surface layer region III, respectively, m; $\Delta_i = \delta_i/h_o$ for parallel plate, $\Delta_i = \delta_i/C$ for partial journal bearing; $i = 1, 2, 3$

$\varepsilon$    Partial journal bearing eccentricity ratio

$\mu$    Core layer (base fluid) viscosity, Ns/m$^2$ ; $\mu = \mu_2$

$\mu_i$    Dynamic viscosity of surface (or porous) layer region I, couple stress fluid film region II, surface layer region III, respectively, Ns/m$^2$; $i = 1, 2, 3$

$\beta_i$    Dynamic viscosity ratio of surface (or porous) layer region I to core layer, dynamic viscosity ratio of surface layer region III to core layer, respectively; $\beta_i = \mu_i/\mu; i = 1, 3$

$\eta$    Couple stress material constant, kgm/s

$\lambda$    Couple stress parameter; $\lambda = (\sqrt{\eta/\mu})/h_o$ for parallel plate, $\lambda = (\sqrt{\eta/\mu})/C$ for partial journal bearing

$\theta$    Angle measured from the center position in partial journal bearing

## 1    Squeeze Film Bearing Lubrication in Synovial Joints

Squeeze film lubrication characteristics for synovial joints identified as bearings are influenced by surface and lubricant properties. In squeeze film lubrication, fluid between two lubricated surfaces approaching each other is squeezed out and pressure is built up to support the applied load. The load capacity mostly relies on the pressure field induced by the squeezed fluid velocity caused by the approaching surfaces. There is an increasing focus on research on tribological mechanisms and lubrication of synovial joints applications. Mattei et al. [1] reviewed literature on lubrication and wear models of hip implants that have been used to investigate the effects of geometric and material parameters. The need for including both lubrication and wear aspects that can help in the development of hip implants is highlighted. Myant and Cann [2] reviewed lubrication models of artificial hip joints and suggested that interfacial film formation is driven by aggregation of synovial fluid proteins in shear flow to high viscosity which is entrained into the contact zone. Boedo and Coots [3] investigated the wear characteristics of a novel squeeze film hip implant design based on geometry coupled with kinematic and load characteristics of the human gait cycle. The squeeze film design employing low modulus elastic elements with high modulus metallic coatings showed a promising alternative to contemporary artificial hip joint designs. Pascau et al. [4] investigated the role of fluid film lubrication in total knee replacement systems during the stance phase of walking under various joint conformity conditions. The hydrodynamic lubrication at the early stance phase, coupled with high conformity, helps to decrease significantly the compressive stresses. The synovial joint lubrication investigations are significant in design of total knee replacement (TKR) with large variations in conformity and total hip replacement (THR) based on significant conformity.

This study aims to explore layered squeeze film parallel plate and partial journal bearing described through porous–surface adsorbent layers with couple stress fluid film core region. The porous–surface adsorbent Newtonian layers are of higher viscosity than couple stress fluid core layer of conventional viscosity.

## 1.1  Squeeze Film Bearings in Synovial Joints

Ruggiero et al. [5, 6] presented synovial fluid film force for squeeze film lubrication of the human ankle joint using closed-form description. The synovial fluid motion through a porous cartilage matrix is depicted as fluid transport across the layered articular cartilage medium [5]. The couple stress characteristics of synovial fluid and porosity of cartilage matrix are considered in the evaluation of fluid film force acting on the synovial human ankle joint during squeeze motion [6]. Mongkolwongrojn et al. [7] presented transient investigation of artificial knee joint in elasto-hydrodynamic lubrication for point contact with non-Newtonian lubricants. Under similar operating conditions, the minimum film thickness obtained for a shear thinning non-Newtonian fluid is lower than a Newtonian fluid. Bujurke and Kudenatti [8] presented squeeze film behavior of poroelastic bearings considering surface roughness of cartilage and couple stress behavior of synovial fluid. Abdullah et al. [9] investigated surface roughness and hydrodynamic lubrication impact on the performance of articular cartilage of synovial human knee joint while gait cycle. Sinha et al. [10] analyzed behavior of synovial fluid lubrication of human joints governed by micro-polar fluid theory. Synovial fluid contains long-chain hyaluronic acid chain molecules causes an increase in the effective viscosity near the porous cartilage, by way of an increase in the micro-motions and the couple stress.

### 1.1.1 Parallel Plate

Lin et al. [11] investigated the performance of two parallel squeeze film plates considering non-Newtonian couple stresses and convective fluid inertia forces. Combined response of surface quality as well as couple stress fluid in squeeze film poroelastic bearings of synovial joints was investigated by Bujurke et al. [12]. The effects of surface quality and elasticity are more prominent for poroelastic bearings with couple stress fluids.

### 1.1.2 Partial Journal Bearing

Walicki and Walicka [13] investigated an enhancement in couple stress fluid squeeze film behavior of a synovial joint using hemispherical bearing. Lin [14] predicted the performance of infinitely long partial journal bearing considering the couple stress fluid squeeze film.

## 1.2 Couple Stress Fluids

The long-chain molecules as polar additives in synovial fluid are characterized as couple stress fluids. Stokes [15] couple stress fluid model derived based on microcontinuum theory is the simplest generalization of the classical theory of fluids. The effect of couple stresses might be expected to appear to a noticeable extent in lubricants containing additives with long-chain molecules. Stokes couple stress fluid model allows for the polar effects such as the presence of couple stresses and body couples. The effects of couple stresses on the squeeze film motion of the synovial joints were analyzed. These investigations on the presence of couple stresses in synovial fluids have led to the enhancement in the load carrying capacity of joints in comparison with the Newtonian lubricant case.

## 1.3 Layered Lubrication Analysis

The additives have proved to improve the performance of thin film hydrodynamic lubricated contacts forming a thin porous layer adhering to bearing surfaces.

### 1.3.1 Porous Layer

Many researchers have considered the effects of porosity [16] of cartilage surfaces of the synovial joints. Li and Chu [17] and Elsharkawy [18] presented the effect of porous adsorbent layers and couple stress fluid core layer for improvement in thin film lubrication performance of journal bearing. To investigate the effects of lubricant additives, the boundary layer microstructure of lubricating surfaces are modeled as thin porous films adsorbed on bearing surfaces. The Brinkman model is used to analyze flow in the porous region and Stokes micro-continuum theory is utilized to model the couple stress effects. Rao et al. [19] presented an analysis of journal bearing with double-layered porous adsorbent film on bearing surfaces with couple stress fluid. The surface layer is simulated as a porous layer with infinite permeability. A low-permeability porous layer interface with core conventional viscosity layer in a lubricant film with double-layered porous layer configuration improves journal bearing performance characteristics.

### 1.3.2  Surface Layer

Tichy [20] derived generalized Reynolds equation using surface layer model wherein, surface layers possessed higher viscosity than core layer which is pertinent to thin film lubrication. With an increase in thickness and viscosity of surface layers, higher load capacity and lower coefficient of friction are obtained for journal bearing. Szeri [21] investigated significant savings in journal bearing power loss due to reduction in viscous friction based on a structure of a composite film that combines high- and low-viscosity fluids. Rao et al. [22] evaluated three-layered film journal bearing for improvement in load capacity and decline in friction coefficient lubricated with nanoparticle mixed couple stress fluid.

## 2  Parallel Plate Layered Lubrication with Couple Stress Fluids

This study investigates adsorbent layered couple stress squeeze film lubrication in parallel plate. Figure 1 shows parallel plate with layered lubrication. An adsorbent layered lubrication with couple stress fluid film core region is described through the

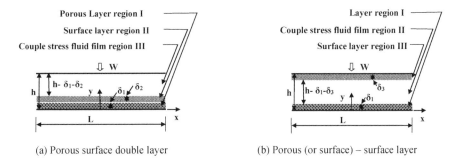

(a) Porous surface double layer                      (b) Porous (or surface) – surface layer

**Fig. 1**  Parallel plate with layered lubrication

following configurations: (i) porous–surface adsorbent double layers (I, II) on stationary plate (Fig. 1a), (ii) porous–surface adsorbent layer (I, III) on stationary–moving
plate (Fig. 1b), and (iii) surface–surface adsorbent layer (I, III) on stationary–moving
plate (Fig. 1b). The fluid viscosity in the porous layer is considered as conventional
core layer viscosity while the surface adsorbent layer viscosity is considered higher
than conventional core layer viscosity. A one-dimensional modified Reynolds equation is derived based on Newtonian fluid model in surface layer region, Brinkman
model in porous layer region and Stokes model in couple stress fluid film region.
The non-dimensional film thickness in layered parallel plate is $H$. The variation of
squeeze film pressure across the layered fluid film is assumed to be negligible. The
thickness of surface (or porous) adsorbent layers is considered as constant. Rao et al.
[23] analyzed the squeeze film load capacity characteristics of three-layered parallel plate lubricated with couple stress fluids for skeletal joint applications using
one-dimensional analysis.

## 2.1  Porous–Surface Double-Layer Parallel Plate

The momentum equations for porous layer (region I: $0 \le y \le \delta_1$), Newtonian surface
layer (region II: $\delta_1 \le y \le \delta_1 + \delta_2$), and the equations of motion for Stokes' couple
stress fluid film (region III: $\delta_1 + \delta_2 \le y \le h$) based on thin film lubrication theory
are

$$\frac{1}{\mu_1}\frac{\mathrm{d}p}{\mathrm{d}x} = -\frac{u_1}{K_1} + \frac{\mathrm{d}^2 u_1}{\mathrm{d}y^2} \tag{1}$$

$$\frac{1}{\mu_2}\frac{\mathrm{d}p}{\mathrm{d}x} = \frac{\mathrm{d}^2 u_2}{\mathrm{d}y^2} \tag{2}$$

$$\frac{1}{\eta}\frac{\mathrm{d}p}{\mathrm{d}x} = \frac{\mu_3}{\eta}\frac{\mathrm{d}^2 u_3}{\mathrm{d}y^2} - \frac{\mathrm{d}^4 u_3}{\mathrm{d}y^4} \tag{3}$$

The boundary conditions for velocity at the stationary plate, at the interface of
stationary plate adsorbent porous and surface layer, at the interface of surface and
core couple stress film, and at moving plate, respectively, are

$$y = 0 : u_1 = 0 \tag{4}$$

$$y = \delta_1 : u_1 = u_2 = u_{12}, \quad \mu_1 \frac{\mathrm{d}u_1}{\mathrm{d}y} = \mu_2 \frac{\mathrm{d}u_2}{\mathrm{d}y} \tag{5}$$

$$y = \delta_1 + \delta_2 : u_2 = u_3 = u_{23}, \quad \mu_2 \frac{\mathrm{d}u_2}{\mathrm{d}y} = \mu_3 \frac{\mathrm{d}u_3}{\mathrm{d}y}, \frac{\mathrm{d}^2 u_3}{\mathrm{d}y^2} = 0 \tag{6}$$

$$y = h : u_3 = 0, \quad \frac{\mathrm{d}^2 u_3}{\mathrm{d}y^2} = 0 \tag{7}$$

Integrating Eq. (1) for porous layer ($0 \leq Y \leq \Delta_1$), Eq. (2) for surface layer ($\Delta_1 \leq Y \leq \Delta_1 + \Delta_2$), and Eq. (3) for couple stress fluid film ($\Delta_1 + \Delta_2 \leq Y \leq H$) using the boundary conditions in (4)–(7), the non-dimensional fluid velocity distribution in porous layer ($0 \leq Y \leq \Delta_1$), surface layer ($\Delta_1 \leq Y \leq \Delta_1 + \Delta_2$), and couple stress fluid film ($\Delta_1 + \Delta_2 \leq Y \leq H$) are expressed as

$$0 \leq Y \leq \Delta_1 : U_1 = U_{12} \frac{\sinh\left(\frac{Y}{\sqrt{K_1}}\right)}{\sinh\left(\frac{\Delta_1}{\sqrt{K_1}}\right)} - K_1 \frac{\mathrm{d}P}{\mathrm{d}X} C_1 \tag{8}$$

$$\Delta_1 \leq Y \leq \Delta_1 + \Delta_2 : U_2 = \frac{1}{2\beta_2} \frac{\mathrm{d}P}{\mathrm{d}X}(Y - \Delta_1)(Y - \Delta_1 - \Delta_2)$$
$$- U_{12}\left(\frac{Y - \Delta_1 - \Delta_2}{\Delta_2}\right) + U_{23}\left(\frac{Y - \Delta_1}{\Delta_2}\right) \tag{9}$$

$$\Delta_1 + \Delta_2 \leq Y \leq H : U_3 = -U_{23}\left(\frac{Y - H}{H - \Delta_1 - \Delta_2}\right)$$
$$+ \frac{1}{2}\frac{\mathrm{d}P}{\mathrm{d}X}(Y - H)(H - \Delta_1 - \Delta_2) + \lambda^2 \frac{\mathrm{d}P}{\mathrm{d}X}C_2 \tag{10}$$

where the coefficients are derived as

$$C_1 = 1 - \frac{\sinh\left(\frac{Y}{\sqrt{K_1}}\right)}{\sinh\left(\frac{\Delta_1}{\sqrt{K_1}}\right)} + \frac{\sinh\left(\frac{Y - \Delta_1}{\sqrt{K_1}}\right)}{\sinh\left(\frac{\Delta_1}{\sqrt{K_1}}\right)},$$

$$C_2 = 1 + \frac{\sinh\left(\frac{Y - H}{\lambda}\right) - \sinh\left(\frac{Y - \Delta_1 - \Delta_2}{\lambda}\right)}{\sinh\left(\frac{H - \Delta_1 - \Delta_2}{\lambda}\right)} \tag{11}$$

$$U_{12} = -\frac{\mathrm{d}P}{\mathrm{d}X}I_1, \quad U_{23} = -\frac{\mathrm{d}P}{\mathrm{d}X}I_2 \tag{12}$$

$$I_1 = \frac{E_{22}E_{13} - E_{12}E_{23}}{E_{11}E_{22} - E_{12}E_{21}}, \quad I_2 = \frac{-E_{21}E_{13} + E_{11}E_{23}}{E_{11}E_{22} - E_{12}E_{21}} \tag{13}$$

$$E_{11} = \frac{1}{\sqrt{K_1}}\coth\left(\frac{\Delta_1}{\sqrt{K_1}}\right) + \frac{\beta_2}{\Delta_2},$$

$$E_{12} = E_{21} = -\frac{\beta_2}{\Delta_2},$$

$$E_{22} = \frac{\beta_2}{\Delta_2} + \frac{1}{(H - \Delta_1 - \Delta_2)},$$

$$E_{13} = H_1^* + \frac{\Delta_2}{2}, \quad E_{23} = -\lambda H^* + \frac{1}{2}(H - \Delta_1)$$

$$H_1^* = \sqrt{K_1}\left[\coth\left(\frac{\Delta_1}{\sqrt{K_1}}\right) - \operatorname{csch}\left(\frac{\Delta_1}{\sqrt{K_1}}\right)\right]$$

$$H^* = \left[\coth\left(\frac{H - \Delta_1 - \Delta_2}{\lambda}\right) - \operatorname{csch}\left(\frac{H - \Delta_1 - \Delta_2}{\lambda}\right)\right] \qquad (14)$$

The continuity equation across the film in non-dimensional form is

$$Q = \int_0^{\Delta_1} U_1 dY + \int_{\Delta_1}^{H - \Delta_3} U_2 dY + \int_{H - \Delta_3}^{H} U_3 dY, \quad Q = -G\frac{dP}{dX} \qquad (15)$$

where the coefficient of continuity across the film yields

$$G = I_1\left(H_1^* + \frac{\Delta_2}{2}\right) + K_1\left(\Delta_1 - 2H_1^*\right) + \frac{1}{2}I_2(H - \Delta_1)$$

$$+ \frac{1}{12}\left((H - \Delta_1 - \Delta_2)^3 + \frac{\Delta_2^3}{\beta_2}\right) - \lambda^2(H - \Delta_1 - \Delta_2) + 2\lambda^3 H^* \qquad (16)$$

For $\Delta_2 = 0$, $G$ in Eq. (16) reduces to

$$G = \frac{E_{13}}{E_{11}}\left(H_1^* + \frac{1}{2}(H - \Delta_1)\right) + K_1\left(\Delta_1 - 2H_1^*\right)$$

$$+ \frac{1}{12}(H - \Delta_1)^3 - \lambda^2(H - \Delta_1) + 2\lambda^3 H^* \qquad (17)$$

where the coefficients are shown in Eq. (18).

$$E_{11} = \frac{1}{\sqrt{K_1}}\coth\left(\frac{\Delta_1}{\sqrt{K_1}}\right) + \frac{1}{(H - \Delta_1)},$$

$$E_{13} = H_1^* - \lambda H^* + \frac{1}{2}(H - \Delta_1)$$

$$H^* = \left[\coth\left(\frac{H - \Delta_1}{\lambda}\right) - \operatorname{csch}\left(\frac{H - \Delta_1}{\lambda}\right)\right] \qquad (18)$$

For $\Delta_1 = 0$, $G$ in Eq. (16) reduces to

$$G = \frac{1}{2}\frac{E_{13}}{E_{11}}H + \frac{1}{12}\left((H - \Delta_2)^3 + \frac{\Delta_2^3}{\beta_2}\right) - \lambda^2(H - \Delta_2) + 2\lambda^3 H^* \qquad (19)$$

where the coefficients are

$$E_{11} = \frac{\beta_2}{\Delta_2} + \frac{1}{(H - \Delta_2)}, \quad E_{13} = -\lambda H^* + \frac{1}{2}H,$$

$$H^* = \left[ \coth\left(\frac{H - \Delta_2}{\lambda}\right) - \operatorname{csch}\left(\frac{H - \Delta_2}{\lambda}\right) \right]$$  (20)

For $\Delta_1 = 0$ and $\Delta_2 = 0$, $G$ in Eq. (16) reduces to

$$G = \frac{1}{12} H^3 - \lambda^2 H + 2\lambda^3 H^*$$  (21)

The non-dimensional modified Reynolds equation for squeeze motion is expressed as

$$-1 + \frac{dQ}{dX} = 0$$  (22)

The squeeze film boundary conditions are

$$\left. \frac{dP}{dX} \right|_{X=0} = 0, \quad P|_{X=\pm 0.5} = 0$$  (23)

Integrating the Eq. (22) and substituting the boundary conditions in Eq. (23), the non-dimensional squeeze film pressure and load capacity in layered parallel plate are derived as

$$P = - \int_{-0.5}^{X} \frac{X}{G} dX, \quad W = \int_{-0.5}^{0.5} P dX$$  (24)

The parameters employed to investigate outcome of non-dimensional squeeze film load capacity ($W$) of porous–surface double-layer parallel plate are: non-dimensional film thickness of parallel plate ($H$) = 0.5–0.9; couple stress parameter ($\lambda$) = 0.0–0.2; porous–surface adsorbent double-layer thickness ratios ($\Delta_1$) = 0.0–0.12; ($\Delta_2$) = 0.08; non-dimensional permeability of porous layer ($K_1$) = $10^{-1}$–$10^{-5}$; and dynamic viscosity ratio of surface to core layer ($\beta_2$) = 10. The impact of non-dimensional thickness of porous layer ($\Delta_1$) and non-dimensional-film thickness of parallel plate ($H$) on the squeeze film non-dimensional load capacity enhancement are analyzed.

Figure 2a, b illustrate the non-dimensional load capacity ($W$) of parallel plate with porous–surface double layer. The non-dimensional load capacity ($W$) increases with increase in the non-dimensional thickness of porous layer ($\Delta_1$) and decrease in the non-dimensional film thickness of parallel plate ($H$). Higher couple stress parameter ($\lambda = 0.2$) and lower non-dimensional permeability ($K_1 = 10^{-5}$) have a significant effect on the increase in non-dimensional load capacity ($W$) compared to higher non-dimensional permeability ($K_1 = 10^{-1}$). The decline in the non-dimensional permeability of porous layer increases resistance for fluid flow in porous layer.

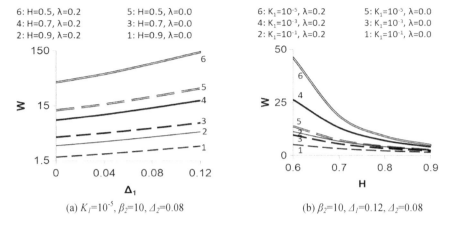

(a) $K_1=10^{-5}$, $\beta_2=10$, $\Delta_2=0.08$                    (b) $\beta_2=10$, $\Delta_1=0.12$, $\Delta_2=0.08$

**Fig. 2** Non-dimensional load capacity of porous surface double-layer parallel plate

## 2.2 Porous–Surface Layer Parallel Plate

The coefficient of continuity across the film in porous– surface layer parallel plate used in determination of non-dimensional squeeze film pressure and load capacity is derived as [22]

$$G = I_1\left(H_1^* + \frac{1}{2}(H - \Delta_1 - \Delta_3)\right) + K_1\left(\Delta_1 - 2H_1^*\right) + \frac{1}{2}I_2(H - \Delta_1)$$
$$+ \frac{1}{12}\left((H - \Delta_1 - \Delta_3)^3 + \frac{\Delta_3^3}{\beta_3}\right) - \lambda^2(H - \Delta_1 - \Delta_3) + 2\lambda^3 H^* \qquad (25)$$

where the coefficients are

$$I_1 = \frac{E_{22}E_{13} - E_{12}E_{23}}{E_{11}E_{22} - E_{12}E_{21}}, \quad I_2 = \frac{-E_{21}E_{13} + E_{11}E_{23}}{E_{11}E_{22} - E_{12}E_{21}} \qquad (26)$$

$$E_{11} = \frac{1}{\sqrt{K_1}}\coth\left(\frac{\Delta_1}{\sqrt{K_1}}\right) + \frac{1}{(H - \Delta_1 - \Delta_3)},$$

$$E_{12} = E_{21} = -\frac{1}{(H - \Delta_1 - \Delta_3)},$$

$$E_{22} = \frac{\beta_3}{\Delta_3} + \frac{1}{(H - \Delta_1 - \Delta_3)},$$

$$E_{13} = H_1^* - \lambda H^* + \frac{1}{2}(H - \Delta_1 - \Delta_3), \quad E_{23} = -\lambda H^* + \frac{1}{2}(H - \Delta_1)$$

$$H_1^* = \sqrt{K_1}\left[\coth\left(\frac{\Delta_1}{\sqrt{K_1}}\right) - \operatorname{csch}\left(\frac{\Delta_1}{\sqrt{K_1}}\right)\right]$$

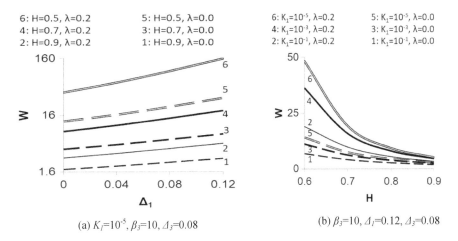

(a) $K_1=10^{-5}$, $\beta_3=10$, $\Delta_3=0.08$                     (b) $\beta_3=10$, $\Delta_1=0.12$, $\Delta_3=0.08$

**Fig. 3** Non-dimensional load capacity of porous–surface layer parallel plate

$$H^* = \left[\coth\left(\frac{H - \Delta_1 - \Delta_3}{\lambda}\right) - \operatorname{csch}\left(\frac{H - \Delta_1 - \Delta_3}{\lambda}\right)\right] \qquad (27)$$

The factors employed to examine results of non-dimensional squeeze film load capacity ($W$) of porous–surface layer parallel plate are: non-dimensional film thickness of parallel plate ($H$) = 0.5–0.9; couple stress parameter ($\lambda$) = 0.0–0.2; porous–surface adsorbent layer thickness ratios ($\Delta_1$) = 0.0–0.12; ($\Delta_3$) = 0.08; non-dimensional permeability of porous layer ($K_1$) = $10^{-1}$–$10^{-5}$; and ratio of dynamic viscosity of surface to core layer ($\beta_3$) = 10. The influence of non-dimensional thickness of porous layer ($\Delta_1$) and non-dimensional film thickness of parallel plate ($H$) on the squeeze film non-dimensional load capacity enhancement are analyzed.

The influence of porous–surface film parallel plate is investigated in Fig. 3a, b. Similar to Fig. 2a, b at lower non-dimensional permeability ($K_1 = 10^{-5}$), the non-dimensional load capacity ($W$) increases with increase in the non-dimensional thickness of porous layer ($\Delta_1$) and decrease in non-dimensional film thickness of parallel plate ($H$) with couple stress fluid core layer ($\lambda = 0.2$). Lower non-dimensional permeability ($K_1 = 10^{-5}$) with couple stress fluid core layer ($\lambda = 0.2$) has a crucial impact on the rise in non-dimensional load capacity ($W$) than higher non-dimensional permeability ($K_1 = 10^{-1}$).

## 2.3  Surface–Surface Layer Parallel Plate

The coefficients of continuity across the film in surface–surface layer parallel plate used in determination of non-dimensional squeeze film pressure and load capacity is derived as [22]

$$G = \frac{1}{2}I_1(H - \Delta_3) + \frac{1}{2}I_2(H - \Delta_1)$$

$$+ \frac{1}{12}\left(\frac{\Delta_1^3}{\beta_1} + (H - \Delta_1 - \Delta_3)^3 + \frac{\Delta_3^3}{\beta_3}\right) - \lambda^2(H - \Delta_1 - \Delta_3) + 2\lambda^3 H^* \quad (28)$$

where the coefficients are

$$E_{11} = \frac{\beta_1}{\Delta_1} + \frac{1}{(H - \Delta_1 - \Delta_3)}, \quad E_{13} = -\lambda H^* + \frac{1}{2}(H - \Delta_3) \quad (29)$$

The coefficients $I_1, I_2, E_{12} = E_{21}, E_{22}, E_{23}, H^*$ described in Eqs. (26)–(27) are considered in the analysis.

The parameters used to analyze results of non-dimensional squeeze film load capacity ($W$) of surface–surface layer parallel plate are: non-dimensional film thickness of parallel plate ($H$) = 0.5–0.9; couple stress parameter ($\lambda$) = 0.0–0.2; surface—surface adsorbent layer thickness ratios ($\Delta_1$) = 0.0–0.12; ($\Delta_3$) = 0.08; and dynamic viscosity ratios of surface to core layers ($\beta_1$) = 2, 5, 10; ($\beta_3$) = 10.

The influence of non-dimensional thickness of surface layer ($\Delta_1$) and non-dimensional film thickness of parallel plate ($H$) on the squeeze film non-dimensional load capacity enhancement are analyzed in Fig. 4a, b. The non-dimensional load capacity ($W$) increases with increasing non-dimensional thickness of surface layer ($\Delta_1 = 0.0$–$0.12$) and decreasing non-dimensional film thickness of parallel plate ($H = 0.9$–$0.5$) for couple stress fluids ($\lambda = 0.2$). The non-dimensional load capacity ($W$) increases with decreasing non-dimensional film thickness of parallel plate ($H$) and increasing dynamic viscosity ratio of surface layer to core layer ($\beta_1 = 2$–$10$). The non-dimensional load capacity ($W$) increases with increasing squeeze film pressure. Augmented non-dimensional load capacity ($W$) is achieved with higher

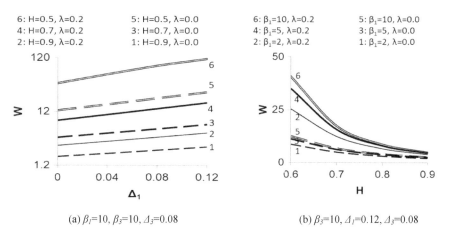

(a) $\beta_1$=10, $\beta_3$=10, $\Delta_3$=0.08                                    (b) $\beta_3$=10, $\Delta_1$=0.12, $\Delta_3$=0.08

**Fig. 4** Non-dimensional load capacity of surface–surface layer parallel plate

non-dimensional surface layer thickness ($\Delta_1 = 0.12$) and lower non-dimensional film thickness of parallel plate ($H = 0.6$) for couple stress fluids ($\lambda = 0.2$).

## 3 Partial Journal Bearing Layered Lubrication with Couple Stress Fluids

This study investigates adsorbent layered couple stress squeeze film lubrication in partial journal bearing. Figure 5 shows an adsorbent layered lubrication in partial journal bearing with couple stress fluid film core region described through the following configurations: (i) porous–surface adsorbent double layers (I, II) on stationary plate (Fig. 5a), (ii) porous–surface adsorbent layer (I, III) on stationary–moving plate (Fig. 5b), and (iii) surface–surface adsorbent layer (I, III) on stationary–moving plate (Fig. 5b). The fluid viscosity in porous layer is considered as conventional core layer viscosity while the surface adsorbent layer viscosity is considered higher than conventional core layer viscosity. The thickness of porous (or surface) adsorbent layers is considered as constant. Squeeze film non-dimensional load capacity of couple stress fluid lubricated layered partial journal bearing is presented. Rao et al. [21] analyzed the squeeze film load capacity characteristics of three-layered partial journal bearing lubricated with couple stress fluids using one-dimensional analysis.

### *3.1  Porous–Surface Double-Layer Partial Journal Bearing*

The non-dimensional squeeze film pressure and load capacity in layered partial journal bearing are obtained as

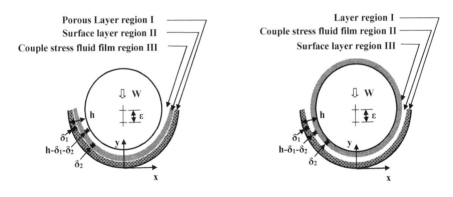

(a) Porous surface double layer                    (b) Porous (or surface) – surface layer

**Fig. 5** Partial journal bearing with layered lubrication

$$P = - \int_{-0.5\pi}^{\theta} \frac{\sin\theta}{G} d\theta, \quad W = \int_{-0.5\pi}^{0.5\pi} P\cos\theta d\theta \qquad (30)$$

The coefficients of continuity across the film derived in Eq. (16) for porous–surface double-layer partial journal bearing is used in the analysis. The non-dimensional film thickness in layered partial journal bearing is $H = 1 - \varepsilon\cos\theta$.

The parameters used to analyze the results of non-dimensional squeeze film load capacity ($W$) of porous–surface double-layer partial journal bearing are: partial journal bearing eccentricity ratio ($\varepsilon$) = 0.1–0.5; couple stress parameter ($\lambda$) = 0.0–0.2; porous–surface adsorbent double-layer thickness ratios ($\Delta_1$) = 0.0–0.12; ($\Delta_2$) = 0.08; non-dimensional permeability of porous layer ($K_1$) = $10^{-1}$–$10^{-5}$; and dynamic viscosity ratio of surface to core layer ($\beta_2$) = 10. The influence of non-dimensional thickness of porous layer ($\Delta_1$) and partial journal bearing eccentricity ratio ($\varepsilon$) on the squeeze film non-dimensional load capacity enhancement are analyzed.

Figure 6a, b show the non-dimensional load capacity ($W$) of partial journal bearing with porous–surface double-layer configuration lubricated with couple stress fluids. The non-dimensional load capacity ($W$) increases with increasing porous adsorbent layer thickness ratio ($\Delta_1$ = 0.0–0.12) and increasing eccentricity ratio ($\varepsilon$ = 0.1–0.5). The resistance to fluid flow in porous layer increases for lower non-dimensional permeability parameter ($K_1$ = $10^{-5}$). Higher partial journal bearing eccentricity ratio ($\varepsilon$ = 0.4) and lower non-dimensional permeability ($K_1$ = $10^{-5}$) show significant influence on increase in non-dimensional load capacity ($W$) with couple stress fluid core layer ($\lambda$ = 0.2).

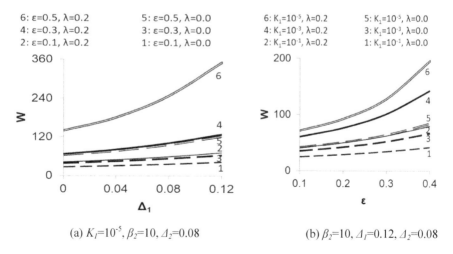

(a) $K_1$=$10^{-5}$, $\beta_2$=10, $\Delta_2$=0.08     (b) $\beta_2$=10, $\Delta_1$=0.12, $\Delta_2$=0.08

**Fig. 6** Non-dimensional load capacity of porous–surface double-layer partial journal bearing

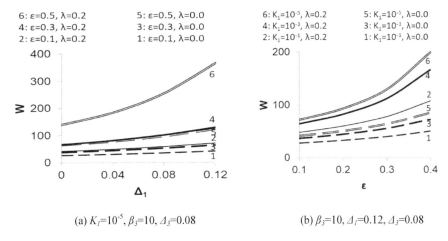

(a) $K_1=10^{-5}$, $\beta_3=10$, $\Delta_3=0.08$                          (b) $\beta_3=10$, $\Delta_1=0.12$, $\Delta_3=0.08$

**Fig. 7** Non-dimensional load capacity of porous–surface layer partial journal bearing

## 3.2  Porous–Surface Layer Partial Journal Bearing

The parameters used to analyze results of non-dimensional squeeze film load capacity ($W$) of porous–surface layer partial journal bearing are: partial journal bearing eccentricity ratio ($\varepsilon$) = 0.1–0.5; couple stress parameter ($\lambda$) = 0.0–0.2; porous–surface adsorbent layer thickness ratios ($\Delta_1$) = 0.0–0.12; ($\Delta_3$) = 0.08; non-dimensional permeability of porous layer ($K_1$) = $10^{-1}$–$10^{-5}$; and dynamic viscosity ratio of surface to core layer ($\beta_3$) = 10. The influence of non-dimensional thickness of porous layer ($\Delta_1$) and partial journal bearing eccentricity ratio ($\varepsilon$) on the squeeze film non-dimensional load capacity enhancement are analyzed.

The influence of porous–surface film partial journal bearing is investigated in Fig. 7a, b. The variation of non-dimensional load capacity ($W$) with increasing non-dimensional thickness of porous layer ($\Delta_1$) and increasing partial journal bearing eccentricity ratio ($\varepsilon$) is presented. The non-dimensional load capacity ($W$) increases with increase in the non-dimensional thickness of porous layer ($\Delta_1 = 0.0$–$0.12$) at lower non-dimensional permeability of porous layer ($K_1 = 10^{-5}$). Higher partial journal bearing eccentricity ratio ($\varepsilon = 0.4$), higher non-dimensional thickness of porous layer ($\Delta_1 = 0.12$), and lower non-dimensional permeability ($K_1 = 10^{-5}$) have significant effect on increase in non-dimensional load capacity ($W$) for couple stress fluids ($\lambda = 0.2$).

## 3.3  Surface–Surface Layer Partial Journal Bearing

The parameters used to analyze results of non-dimensional squeeze film load capacity ($W$) of surface–surface layer partial journal bearing are: partial journal bearing

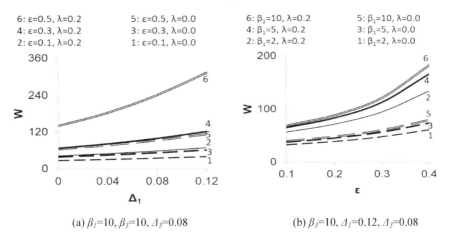

(a) $\beta_1$=10, $\beta_3$=10, $\Delta_3$=0.08                    (b) $\beta_3$=10, $\Delta_1$=0.12, $\Delta_3$=0.08

**Fig. 8** Non-dimensional load capacity of surface–surface layer partial journal bearing

eccentricity ratio $(\varepsilon) = 0.1$–$0.5$; couple stress parameter $(\lambda) = 0.0$–$0.2$; surface–surface adsorbent layer thickness ratios $(\Delta_1) = 0.0$–$0.12$; $(\Delta_3) = 0.08$; and dynamic viscosity ratios of surface to core layers $(\beta_1) = 2, 5, 10$; $(\beta_3) = 10$.

The influence of non-dimensional thickness of surface layer $(\Delta_1)$ and partial journal bearing eccentricity ratio $(\varepsilon)$ on the squeeze film non-dimensional load capacity $(W)$ enhancement are analyzed in Fig. 8a, b. The non-dimensional load capacity $(W)$ increases with increase in non-dimensional thickness of surface layer $(\Delta_1 = 0.0$–$0.12)$ for couple stress fluids $(\lambda = 0.2)$. The non-dimensional load capacity $(W)$ increases with increasing partial journal bearing eccentricity ratio $(\varepsilon = 0.1$–$0.4)$ for couple stress fluids $(\lambda = 0.2)$ with increasing dynamic viscosity ratio of surface layer to core layer $(\beta_1 = 2$–$10)$. Higher non-dimensional load capacity $(W)$ is obtained for couple stress fluids $(\lambda = 0.2)$ with higher non-dimensional thickness of surface layer $(\Delta_1 = 0.12)$ and higher partial journal bearing eccentricity ratio $(\varepsilon = 0.5)$.

## 4  Conclusions

An improvement in squeeze film load capacity of layered parallel plate and partial journal bearing lubricated with couple stress fluids is evaluated using (i) porous–surface double layer, (ii) porous–surface layer, and (iii) surface–surface layer. Adsorbent porous and surface layers are analyzed using permeability and viscosity ratio (surface to core layer) parameters, respectively. Modified Reynolds equation is derived for wide-layered parallel plate and long-layered partial journal bearing based on one-dimensional analysis. Squeeze film boundary conditions are used to solve modified Reynolds equation to determine squeeze film pressure.

The non-dimensional squeeze film load capacity ($W$) increases significantly with lower non-dimensional permeability of porous layer ($K_1 = 10^{-5}$), higher non-dimensional porous–surface layer thickness ($\Delta_1 = 0.12$), higher dynamic viscosity ratio of surface layer to core layer conventional viscosity ($\beta_1$, $\beta_2$, $\beta_3 = 10$), and higher couple stress parameter for core fluid layer ($\lambda = 0.2$).

Higher viscosity and thickness of adsorbent low-permeable porous or surface layers with core couple stress fluid increases the squeeze film load capacity of layered parallel plate and partial journal bearing. Couple stress fluids with adsorbent porous–surface layers enhances load capacity of squeeze film bearings for skeletal joint applications.

# References

1. Mattei L, Puccio FD, Piccigallo B, Ciulli E (2011) Lubrication and wear modelling of artificial hip joints: a review. Tribol Int 44:532–549
2. Myant C, Cann P (2014) On the matter of synovial fluid lubrication: implications for metal-on-metal hip tribology. J Mech Behav Biomed Mater 34:338–348
3. Boedo S, Coots SA (2017) Wear characteristics of conventional and squeeze-film artificial hip joints. J Tribol 139(3):031603
4. Pascau A, Guardia B, Puertolas JA, Gomez-Barrena E (2019) Knee model of hydrodynamic lubrication during the gait cycle and the influence of prosthetic joint conformity. J Orthop Sci 14:68–75
5. Ruggiero A, Gomez E, D'Amato R (2011) Approximate analytical model for the squeeze-film lubrication of the human ankle joint with synovial fluid filtrated by articular cartilage. Tribol Lett 41:337–343
6. Ruggiero A, Gomez E, D'Amato R (2013) Approximate closed form solution of the synovial fluid film force in the human ankle joint with non-Newtonian lubricant. Tribol Int 57:156–161
7. Mongkolwongrojn M, Wongseedakaew K, Kennedy FE (2010) Transient elastohydrodynamic lubrication artificial knee joint with non-Newtonian fluids. Tribol Int 43:1017–1026
8. Bujurke NM, Kudenatti RB (2006) An analysis of rough poroelastic bearings with reference to lubrication mechanism of synovial joints. Appl Math Comp 178:309–320
9. Abdullah EY, Edan NM, Kadhim AN (2017) Study surface roughness and friction of synovial human knee joint with using mathematical model, Special issue: 1st Scientific International Conference, College of Science, Al-Nahrain University, 21st–22th November 2017, Part I, pp 109–118
10. Sinha P, Singh C, Prasad KR (1982) Lubrication of human joints—a micro continuum approach. Wear 80:159–181
11. Lin JR, Hung CR, Lu RF (2006) Averaged inertia principle for non-newtonian squeeze films in wide parallel plates: couple stress fluid. J Marine Sci Techonol 14:218–224
12. Bujurke NM, Kudenatti RB, Awati VB (2007) Effect of surface roughness on squeeze film poroelastic bearings with special reference to synovial joints. Math Biosci 209:76–89
13. Walicki E, Walicka A (2000) Mathematical modelling of some biological bearings. Smart Mater Struct 9:280–283
14. Lin JR (1997) Squeeze film characteristics of long partial journal bearings lubricated with couple stress fluids. Tribol Int 30:53–58
15. Stokes VK (1966) Couple stresses in fluids. Phys Fluids 9:1709–1715
16. Bhui AS, Singh G, Sidhu SS, Bains PS (2018) Experimental investigation of optimal ED machining parameters for Ti-6Al-4 V biomaterial. FU Mech Eng 16(3):337–345

17. Li WL, Chu HM (2004) Modified reynolds equation for couple stress fluids—a porous media model. Acta Mech 171:189–202
18. Elsharkawy AA (2005) Effects of lubricant additives on the performance of hydro dynamically lubricated journal bearings. Tribol Lett 18:63–73
19. Rao TVVLN, Rani AMA, Nagarajan T, Hashim FM (2013) Analysis of journal bearing with double-layer porous lubricant film: Influence of surface porous layer configuration. Tribol Trans 56:841–847
20. Tichy JA (1995) A surface layer model for thin film lubrication. Tribol Trans 38(3):577–582
21. Szeri AZ (2010) Composite-film hydrodynamic bearings. Int J Eng Sci 48:1622–1632
22. Rao TVVLN, Sufian S, Mohamed NM (2013) Analysis of nanoparticle additive couple stress fluids in three-layered journal bearing. J Phys Conf Ser 431:012023
23. Rao TVVLN, Rani AMA, Manivasagam G (2018) Squeeze film analysis of three-layered parallel plate and partial journal bearing lubricated with couple stress fluids for skeletal joint applications. Mater Today Proc (Accepted)

# Chapter 6
# Passive Prosthetic Ankle and Foot with Glass Fiber Reinforced Plastic: Biomechanical Design, Simulation, and Optimization

**Thanh-Phong Dao and Ngoc Le Chau**

## 1 Introduction

Nowadays, amputees are facing difficult challenges during work, activity, life, and transportation. Alongside an increase in the number of amputees, prostheses or orthotics devices are also developing to meet the support of movement. In the USA, there are about two million amputees' loss at least a lower limb or upper limb due to different accidences [1]. Regarding lower limbs under knee, the current work enters into the development of prosthetic ankle–foot so as to support for amputees. In the past decades, many researchers have developed the lower limbs, such as human robots, soft robots, and rehabilitation devices, with an assist of power. In order to work effectively, these robots must bring rigid links, motors, actuators, and a little complicated manufacture which lead a more cost and heavyweights. Many characteristics of a prosthetic ankle–foot have attended. First of all, the strain energy (SE) is desired so as to make comfort gaits. Especially, walking on a roughness surfaces, SE of the prostheses generates movements without external actuators. Therefore, a concept of energy storage and return (ESRA) which speeded up a faster walking was proposed [2–4]. The concept of ESRA has allowed a large stroke and strain energy [5–9]. Along with a large ESRA, structural behaviors and biomechanical performances are very sensitive to material's behavior. How a smart material is suitable for

T.-P. Dao (✉)
Division of Computational Mechatronics, Institute for Computational Science, Ton Duc Thang University, Ho Chi Minh City, Vietnam
e-mail: daothanhphong@tdtu.edu.vn

Faculty of Electrical & Electronics Engineering, Ton Duc Thang University, Ho Chi Minh City, Vietnam

N. Le Chau
Faculty of Mechanical Engineering, Industrial University of Ho Chi Minh City, Ho Chi Minh City, Vietnam

© Springer Nature Singapore Pte Ltd. 2019
P. S. Bains et al. (eds.), *Biomaterials in Orthopaedics and Bone Regeneration*,
Materials Horizons: From Nature to Nanomaterials,
https://doi.org/10.1007/978-981-13-9977-0_6

73

the prosthetic ankle–foot. In improving the biomechanical performance, carbon fiber was used to manufacture the ESAR feet [10, 11]. The results showed the proposed feet gain a good ESRA and provide a support for walking phases [12, 13].

Another aspect of prosthetic ankle–foot is concerns to suppress external vibrations for amputees. During walking, the free and forced vibrations from ground may re-injure to the human body of the amputees. Such vibrations are poor characteristics of the prostheses. Bashar and Jumaa discovered the influences of possible vibrations to the prosthetic feet [14]. Recent years, Cherelle et al. used more flexible actuators so as to enhance the performance efficiency for the prosthetic feet and to make a comfort for walking [15–18]. Starker et al. [19] investigated a prosthetic lower limb for comfortable walking on uphill and downhill. Ko et al. [20] entered into statics and dynamics of lower limbs. An overview of the prosthetic feet can be found in Refs. [21–28]. It is known that two common different types of ankle prostheses in terms of passive and active movements. Passive prostheses are very affordable for amputee rehabilitation. Active prostheses alongside developed rapidly, but it needs power to produce energy. However, the same procedure of both types is how an amputee can mimic the functions of normal gait. Passive devices can include springs to guarantee the ESAR and dampers to suppress vibrations. In the past, a solid ankle-based cushioned heel was considered [29–33]. Active prostheses, called as powered prostheses, include motors and actuators [34–36], and they have a heavyweight for worn lower limbs. Four criteria should be considered to compare the passive prostheses with the active prostheses: (1) Capacity of portability is very important. To complete this purpose, a miniature size and a lightweight are priority criteria. Although the active prostheses have several advantages, they need actuators and motors; therefore, they are limited for the portability. (2) How to remain simple and affordable prostheses. (3) How to supply an efficient power to recharge. (4) A good prosthesis design whose energy is efficient. Criteria 1 and 2 are considered in passive device due to the capacity of portability. Meanwhile, the two remaining criteria are mainly aimed at active devices. In this chapter, we concentrate on the passive type.

Most previous studies used the robots, exoskeletons, and prostheses for lost lower limbs. Although these devices can be controlled but complicated, their weights are difficult for portability. Nowadays, a lightweight structure is a relatively challenge for investigators. Unlike previous studies, the current study develops a lightweight and monolithic structure for prosthetic ankle and foot 1.0 and 2.0. We use the compliant mechanism (CM), called as flexure-based mechanism, for designing both types of 1.0 and 2.0. CM, a special type of mechanical engineering, has a free friction, no backlash, and monolithic fabrication. Hence, total weight of CMs is lightly. Many researchers have taken into account applications of CM [37]. Recently, our group focused on innovation, design, modeling, and optimization for multiple ultrahigh positioning applications [38, 39], a one degree-of-freedom (DOF) mechanism [40], a two DOF mechanism [41], a displacement sensor [42], a two DOF flexure-based mechanism [43], and so on. CMs can be found in various applications, such as aerospace, biomechanics, and soft robotics. An especial application of CMs is a vibration energy harvester [44, 45]. Most of the previous studies have still remained a complicated structure. In particular, there is a lack of ankle–foot creating a compliant ankle joint as natural human ankle. Unlike the previous studies, we propose two novel types of prosthetic ankle and foot, 1.0 and 2.0, to allow the smooth motions and harvest the undesired vibrations. Based on the highlights and emergences of CMs on the elastic deformation, monolithic manufacturing, and vibration harvester, the present work enters into new entries: (1) A new shape an topology of prosthetic ankle–foot is developed using CMs; (2) glass fiber reinforced plastic is used to manufacture the prosthetics; (3) finite element simulation is conducted to describe; and (4) optimization is adopted to improve the strain energy.

This chapter provides a biomechanical design, simulations, and optimization for novel types of prosthetic ankle and foot. Both designs can mimic an actual human foot. The material of glass fiber reinforced plastic is chosen for prototype types. The static and dynamic behaviors of the ankle–foot 1.0 and 2.0 are analyzed through finite element method. Numerical data are collected by using the Taguchi method and finite element analysis. Mathematical model is built via the response surface methodology. The strain energy is then optimized through the differential evolution algorithm.

## 2  Biomechanical Design

### 2.1  Design Criteria

In this chapter, we propose two types, 1.0 and 2.0, of passive prosthetic ankle and foot. Both designs are expected to fit a truly ankle–foot. Following five factors are taken into account during design process.

i.  Materials can be used, such as wood, plastic, and foam for individuals. Meanwhile, carbon fiber meets the functions for shock absorption, energy efficiency, and lightweight in terms of various speeds, running, climbing, descending stairs with a secure, and confident stride. Therefore, glass fiber reinforced plastic is chosen as material for the ankle–foot in this chapter.

ii.  A prosthetic foot has to provide comfort for activities. Comfort allows a more active function of prosthetic ankle–foot.

iii.  A prosthetic ankle–foot should mimic a human foot in terms of various functions, such as walking, running on different surfaces.

iv.  A prosthetic foot should have a multi-axial motion. Prosthetic ankle mimics the natural ankle to allow the foot to move in multiple planes and rotation around an axis. This permits various activities comfortably and confidently on uneven ground.

v.  Prosthetic ankle–foot must have capacity to harvest energy storage as a spring during walking/running phases.

## 2.2 Structural Design

Regarding a biomechanical design, a novel passive prosthetic ankle–foot must be comfortable and more flexible to be similar to a nature of real ankle–foot. Therefore, we designed two models of curvilinear shape for the prosthetic ankle–foot. The principle of design consisted of main steps: (1) The shank is conjuncted to the ankle. (2) The ankle joint is then drawn monolithically with the instep and heel. (3) The toe is also monolithically integrated with Thenar. (4) All these components can be manufactured by 3D printing. In order to apply for amputees, the shank will be assembled to the socket. Figure 1 illustrates the first type 1.0. It is constructed based on the CMs. The profile of curved beam is used to avoid a high stress concentration. The ankle joint can rotate in multiple DOFs and mimic a real gait. The goal of ankle-inspired foot 1.0 is to achieve a good ESAR during walking. The ESAR provided strain energy to push off the foot. Our ankle–foot 1.0 is a passive structure for rehabilitation for amputees. It also includes all elements of a real ankle–foot.

Figure 2 describes the second type 2.0. It is an innovation version of the first type 1.0 by using a curvilinear shape. We designed three gutters on the instep and Thenar so as to permit more flexibility during walking. It includes six DOFs in terms of three translation and three rotations according to the $x, y, z$ axes, as shown in Fig. 3. The first motion is a translation along the $y$-axis. The second motion is a translation along the $z$-axis. The third motion is a translation along the $x$-axis. It is considered as surging state so as to balance on the front of the foot. The fourth motion is a rotation of the ankle about the $y$-axis. It is regarded as a yaw motion so as to regain an alignment after finishing the tilting. The fifth motion is a rotation around the $z$-axis. The sixth motion is a turning around the $x$-axis and the $z$-axis. This state is regarded as a pitch motion, such as ballet dancers.

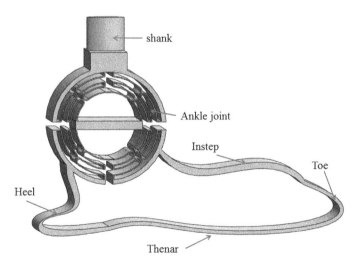

**Fig. 1** A novel 3D compliant prosthetic ankle and foot 1.0

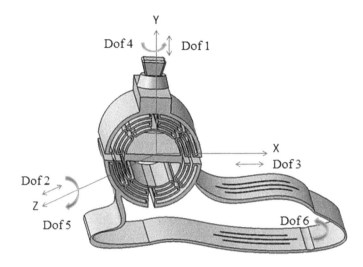

**Fig. 2**   A novel 3D compliant prosthetic ankle and foot 2.0

**Fig. 3**   Human gait cycle

It is better to select the version 2.0 because it provides flexibility and more compliance for the amputees. Another consideration of design is to mimic the human gait. Basically, a gait cycle can be separated into three key phases, as illustrated in Fig. 3. The first phase is heel strike, the second phase is midstance, and the last phase is toe-off. Table 1 shows controllable factors of simulations of the ankle–foot 1.0 and 2.0.

**Table 1**   Parameters of the ankle–foot (Unit: mm)

| Name | Symbol | Values |
|---|---|---|
| Radius of ankle 1.0 & 2.0 | $r_a$ | 60 |
| Width of foot 1.0 & 2.0 | $w_f$ | 160 |
| Length of foot 1.0 & 2.0 | $l_f$ | 320 |
| Thickness of thenar 1.0 & 2.0 | $h_{th}$ | 4 |
| Thickness of heel 1.0 & 2.0 | $h_h$ | 1.5 |
| Thickness of toe 1.0 & 2.0 | $h_t$ | 4 |
| Thickness of instep 1.0 & 2.0 | $h_i$ | 6.5 |
| Width of gutters only 2.0 | $b_g$ | 2 |

# 3 Simulations

This section presents simulations through ANSYS 16.0 software. We applied finite element analysis (FEA) to monitor the static and dynamic behaviors of the prosthetics 1.0 and 2.0. Firstly, 3D models of both types 1.0 and 2.0 are designed by SolidWorks. Figure 4a shows the meshed model for both models by using automatic meshing. Regarding the version 1.0, the mesh consists of 49,769 nodes and 24,197 elements while the version 2.0 includes 39,520 nodes and 19,474 elements, as given in Fig. 4b. In order to manufacture the both types of prosthetics, although there are several popular materials, such as steel, PE, or ABS, glass fiber reinforced plastic (GFRP) is utilized for the simulations and manufacturing because of its high yield strength. The properties of this material are given as in Table 2.

In order to monitor the biomechanical behaviors of the ankle–foot, the version 1.0 is chosen for this analysis. Boundary conditions for the simulations are in terms of heel strike and midstance, and toe-off phase. Loads include a reaction force from the ground, and a moment is exerted to the ankle–foot, shown in Fig. 5a–c.

It is known that the ankle and toe are the main parts for a prosthetic foot. Therefore, we apply a load of 1000 N to the toe-off; the strain energy and total deformation of version 1.0 are retrieved, as given in Fig. 6a, b. However, we found that the strain energy was still low, and therefore, we optimized this energy for the version 2.0.

**(a)**                                          **(b)**

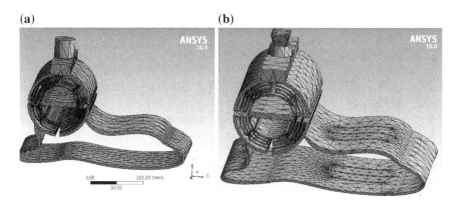

**Fig. 4** Meshing of the prosthetic ankle–foot: **a** version 1.0 and **b** version 2.0

| **Table 2** Mechanical properties of GFRP material | Density | Young's modulus | Poisson's ratio | Yield strength |
|---|---|---|---|---|
| | 1760 kg/m$^3$ | 130 MPa | 0.34 | 148 MPa |

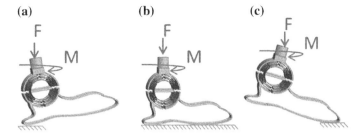

**Fig. 5** Simulations with three phases: **a** heel, **b** midstance, **c** toe

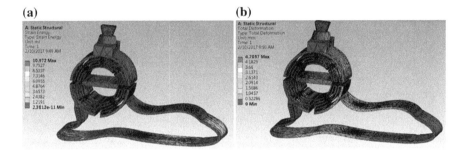

**Fig. 6** Distribution of prosthetic version 1.0: **a** strain energy and **b** total deformation

## 3.1 Stiffness

This section presents the mathematical model and FEA simulations for the ankle–foot 1.0. Similarly, the version is computed by this way as well.

The stiffness of the ankle–foot 1.0 is determined by:

$$K = F/d \tag{1}$$

where $F$ represents the load and $d$ is the total displacement. Meanwhile, a moment is exerted, the stiffness of the ankle–foot is calculated by:

$$K = M/\theta \tag{2}$$

The effects of load on the deflection distribution are analyzed through simulations. In order to conduct this analysis, $F$ is applied from 10 to 100 N, and the plot of the displacement versus the load is given in Fig. 7. The result found the stiffness of the ankle–foot in terms on midstance, and heel is $K_M = K_H = 40$ N/mm. Meanwhile, the stiffness of the ankle–foot in terms of toe-off is approximately $K_t$ of 17 N/mm. The results noted the maximum displacement is in terms of the boundary condition for the toe. However, total deformation is almost the same in terms of boundary conditions of midstance and the heel, respectively.

Fig. 7 Diagram of
displacement versus load of
prosthetic version 1.0

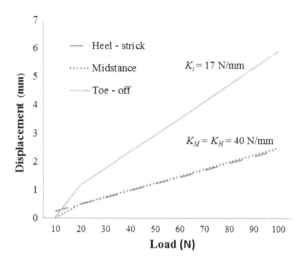

## 3.2   Reaction Moment

This section describes the rotation motion for the ankle–foot 1.0. We expect to make
a prosthetic ankle with multiple degrees of freedom. Hence, we apply a torque in the
range from 100 to 1000 Nmm for the further FEA simulations. As plotted in Fig. 8,
the results found the maximum displacement is in terms of the heel trick and toe-off,
but the lowest value is at boundary condition of the midstance.

Fig. 8 Diagram of
displacement versus moment
of prosthetic version 1.0

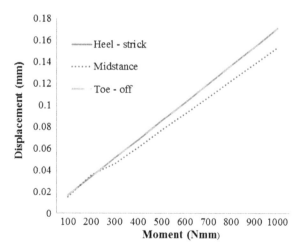

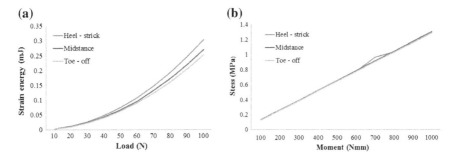

**Fig. 9** Diagram prosthetic version 1.0: **a** strain energy versus force and **b** strain energy versus moment

## 3.3  Strain Energy

This section identifies the strain energy for the ankle–foot 1.0. It is known that each of walking phase, the foot is subjected a reaction force from the ground. Hence, a load from 10 to 100 N, considered as ground reaction forces are exerted. And then, a torque in the range from 100 to 1000 Nmm is applied to the ankle. As seen in Fig. 9a, b, the results noted the energy strain of 0.3 mJ is maximum at the instep. However, the strain energy is maximum in terms of the midstance with applied moment, respectively.

## 3.4  Stress Analysis

This section presents a distribution of stress by FEA simulations for version 1.0. The range of values of forces and moments is pre-used as previous steps. As depicted in Fig. 10, the results found the stress of 28.629 MPa is maximum corresponding to boundary condition. It guarantees a safety for the prosthetic ankle–foot without static failures. The ankle–foot is expected to support for a human body of 100 kg.

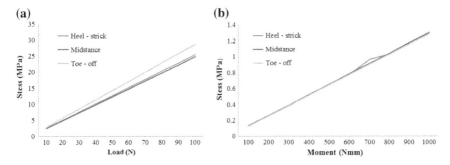

**Fig. 10** Static result of prosthetic version 1.0: **a** stress versus force and **b** stress versus moment

Assuming that a half of body is 50 kg, and therefore, a load 500 N is exerted to the ankle–foot. The result noted the stress of 127 MPa is maximum but less than the yield strength of material (148 MPa). In order to discover the capacity of the ankle–foot, a load up of 600 N is applied. The results found the responding stress of 152 MPa. It is therefore a plastic failure of the ankle–foot. To sum up, the ankle–foot can only support for the weight of 100 kg. A moment is applied to the ankle joint in the range from 100 Nmm to 1000 Nm. The results showed the resulting stress is less than 1.5 MPa, as shown in Fig. 10b. It guarantees a long working time and long fatigue life for the ankle–foot.

## 3.5 Dynamics

During the walking, the ankle–foot is subjected to many external vibrations and it causes a resonant state. This resonance can react to the body. Therefore, this section gives a dynamic simulation for ankle–foot 1.0. The results noted the resonant frequency of first mode is about 30.749 Hz, as given in Fig. 11. Two procedures can be applied to avoid the resonant vibrations: (1) improvement of the first natural frequency or (2) lower the first natural frequency for the structure, but the first way is better selection.

**Fig. 11** Resonant frequencies of the prosthetic version 1.0

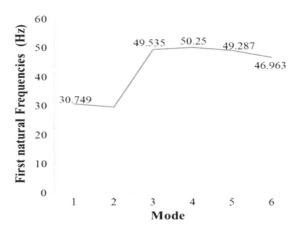

# 4  Optimization

## 4.1  Formulation for Optimization Problem

The goal of this section is to improve the strain energy. As mentioned, the prosthetic version 2.0 is more flexible compared to the version 1.0. Therefore, the version is chosen and then the strain energy is optimized. This energy is expected to be as large as possible. The following optimization problem is briefly described as:

Find the design variables $T_1$, $t_T$, $t_{Th}$, and $t_H$.

Maximize the strain energy:

$$E(t_1, t_T, t_{Th}, t_H) \tag{3}$$

S.t.:

$$\begin{cases} 4\,\text{mm} \leq t_1 \leq 6\,\text{mm} \\ 2\,\text{mm} \leq t_T \leq 4\,\text{mm} \\ 3\,\text{mm} \leq t_{Th} \leq 5\,\text{mm} \\ 1\,\text{mm} \leq t_H \leq 3\,\text{mm} \end{cases} \tag{4}$$

where $E$ is the strain energy. $T_1$, $t_T$, $t_{Th}$, and $t_H$ are the design variables.

## 4.2  Methodology

In order to optimize the strain energy for version 2.0, the numbers of numerical data are layout by using the Taguchi method. Next, the numerical data are got from FEA simulations. Next, the mathematical equation for strain energy is formed through the response surface methodology. Based on the well-formed mathematical model, the strain energy is optimized through differential evolutionary algorithm. A flowchart for optimization process of the strain energy is plotted, as in Fig. 12.

First of all, the Taguchi method [46–48] is used to make the number of numerical experiments. The FEA simulations are applied to retrieve the numerical data. Based on the data, the regression equation of the strain energy for version 2.0 is formed [49, 50] by:

$$E = \beta_0 + \sum_{u=1}^{N} \beta_u x_u + \sum_{u=1}^{N} \beta_{uu} x_u^2 + \sum_{u} \sum_{v} \beta_{uv} x_u x_v + \varepsilon \tag{5}$$

where $E$ represents the output response, $x$ is the inputs, $N$ is the number of design variables, $\beta_u (u = 0, 1, 2, \ldots, N)$ are regression coefficients, $\beta_{uu}$ and $\beta_{uv}$ are quadratic coefficients, and $\varepsilon$ is the model error.

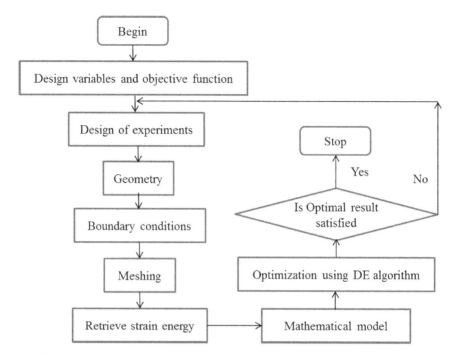

**Fig. 12** Flowchart for optimization process for prosthetic version 2.0

Based on the well-formed regression equation, the strain energy is optimized through the differential evolution (DE) algorithm. The DE includes three phases: (1) mutation, (2) crossover, and (3) selection. The detail of this algorithm can be found in Ref. [49]. The following steps of the DE algorithm are as follows.

Step 1: Initialization
It includes the scale factor $F$, the crossover rate Cr, the population size $M$, and the maximum number of iterations $K$.

Step 2: Candidates
From a uniform distribution in the range $[x_j, x_{jU}](j = 1, 2, \ldots, N)$, where $N$ is the number of variables, the initial candidate solutions are randomly generated.

Step 3: Mutation operation
The mutation operation not only increases the diversity of the solution vectors, but it also increases the exploration capability of the solution space for the algorithm. For each parent $x_i^k (i = 1, 2, \ldots, M)$ of generation $k(k = 1, 2, \ldots, K)$, a trial vector $v_i^{k+1}$ is created by mutating a target vector. Based on the mutation operation, the trial vector can be calculated by:

$$v_i^{k+1} = x_{i_s}^k + F\left(x_{i1}^k + x_{i2}^k\right) \tag{6}$$

Step 4: Crossover operation
This algorithm follows a discrete recombination approach, where elements from the parent vector $x_i^k$ are combined with elements from the trial vector $k_i$ to produce the offspring $u_i^k$:

$$u_{i,j}^{k+1} = \begin{cases} v_{i,j}^{k+1}, & \text{if } r \text{ and } > \text{Cr or } j = r_{1-D} \\ x_{i,j}^k, & \text{otherwise} \end{cases} \tag{7}$$

where $r_{1\sim D}$ is a random integer in $[1, D]$; Cr denotes a crossover rate; and $\text{Cr} \in [0, 1]$.
Step 5: Selection: The offspring $u_i^{k+1}$ from Step 4 replaces the parent $x_i^k$ only if the fitness of the offspring is better than that of the parent.
Step 6: Stopping criterion: If the stopping criterion (maximum number of iterations $K$) is satisfied, computation is terminated. Otherwise, Steps 3 to 5 are repeated.

## 4.3   Results and Discussion

Regarding the prosthetic ankle–foot 2.0, four design variables and they are divided into three levels according to expert knowledge and designer experience, as given in Table 3. $L_9$ orthogonal array is used to layout the number of numerical experiments. Nine numerical experiments are carried out by simulations. Boundary conditions and load are set up in the previous step (Sect. 3). Numerical data are collected, as given in Table 4.

From the numerical data in Table 4, the regression equation for strain energy is found as:

$$E = 550.329 - 23.687t_I + 7.634t_T - 213.280t_{Th} - 7.901t_H$$
$$- 20.740t_It_T + 23.594t_It_{Th} + 23.728t_Tt_{Th} \tag{8}$$

The results found the coefficient of determination for Eq. (8) is $R^2$ of 99.53%. It means that developed equation for the strain energy is robustness. Using Eq. (8), the DE is applied to solve the optimization problem. The results found the optimal parameters are at $t_I = 5.9$ mm, $t_T = 2.0$ mm, $t_{Th} = 3.0$ mm, and $t_H = 1.0$ mm. The

**Table 3** Factors and their levels (Unit: mm)

| Factor | Levels | | |
|---|---|---|---|
| | 1 | 2 | 3 |
| $t_I$: Thickness of inset | 4 | 5 | 6 |
| $t_T$: Thickness of toe | 2 | 3 | 4 |
| $t_{Th}$: Thickness of thenar | 3 | 4 | 5 |
| $t_H$: Thickness of heel | 1 | 2 | 3 |

**Table 4** Numerical results for the strain energy

| No. | $t_I$ (mm) | $t_T$ (mm) | $t_{Th}$ (mm) | $t_H$ (mm) | $E$ (mJ) |
|-----|-----------|-----------|--------------|-----------|----------|
| 1 | 4 | 2 | 3 | 1 | 83.046 |
| 2 | 4 | 3 | 4 | 2 | 20.008 |
| 3 | 4 | 4 | 5 | 3 | 10.972 |
| 4 | 5 | 2 | 4 | 3 | 26.096 |
| 5 | 5 | 3 | 5 | 1 | 16.616 |
| 6 | 5 | 4 | 3 | 2 | 32.085 |
| 7 | 6 | 2 | 5 | 2 | 36.764 |
| 8 | 6 | 3 | 3 | 3 | 31.760 |
| 9 | 6 | 4 | 4 | 1 | 25.134 |

**Table 5** Evaluation of the optimal solutions

| Prosthetic ankle–foot 2.0 | Design variables (mm) | Strain energy (mJ) | Improvement (%) |
|---------------------------|----------------------|--------------------|-----------------|
| Initial design | $t_I = 6.0$, $t_T = 2.0$ $t_{Th} = 5.0$, $t_H = 2.0$ | 36.764 | 155 |
| Optimal design | $t_I = 5.9$, $t_T = 2.0$ $t_{Th} = 3.0$, $t_H = 1.0$ | 93.914 | |

optimal strain energy is about 93.914 mJ, but this value is relatively low. In order to meet a real requirement, the energy strain can be improved by varying the structure, shape, profile, and material. As seen in Table 5, the results also found the strain energy is approximately enhanced 155% compared to the initial design.

## 5 Conclusions

In this chapter, the design, simulations, and optimization for the passive prosthetic ankle–foot 1.0 and 2.0 are presented. The following conclusions are drawn:

- Two models of passive prosthetic ankle–foots are drawn. Two novel types of prosthetic ankle–foot 1.0 and 2.0 are constructed by combination of the bioengineering and compliant mechanism. These prosthetic foots have a very flexible ankle due to the use of CMs. The prosthetic ankle can mimic an actual gait.
- The static and dynamic behaviors of the prosthetic version 1.0 were tested by FEA simulations in terms of various boundary conditions.
- The prosthetic version 2.0 is a more flexibility design compared to the prosthetic ankle–foot 1.0. The version 2.0 is selected for further optimization process.
- The strain energy of version 2.0 is optimized via using the DE algorithm. The strain energy is enhanced about 155% compared to the initial design.

- The results proved the novel prosthetic versions 1.0 and 2.0 can support the body's weight around 100 kg.

# References

1. Grabowski A, D'Andrea S, Herr H (2011) Bionic leg prosthesis emulates biological ankle joint during walking. Proc Ann Meeting Am Soc Biomech, 1–2
2. Casillas JM, Dulieu V, Cohen M, Marcer I, Didier JP (1995) Bioenergetic comparison of a new energy-storing foot and SACH foot in traumatic below-knee vascular amputations. Arch Phys Med Rehabil 76(1):39–44
3. Rao SS, Boyd LA, Mulroy SJ, Bontrager EL, Gronley J, Perry J (1998) Segment velocities in normal and tarsotibial amputees: prosthetic design implications. Trans Rehabil Eng 6(2):219–226
4. Torburn L, Perry J, Ayyappa E, Shanfield SL (1990) Below-knee amputee gait with dynamic elastic response prosthetic feet: a pilot study. J Rehabil Res Dev 27(4):369
5. Lehmann J, Price R, Boswell-Bessette S, Dralle A, Questad K, DeLateur B (1993) Comprehensive analysis of energy storing prosthetic feet: flex foot and seattle foot versus standard SACH foot. Arch Phys Med Rehabil 74(11):1225–1231
6. Macfarlane PA, Nielsen DH, Shurr DG, Meier K (1991) Gait comparisons for below-knee amputees using a flex-foottm versus a conventional prosthetic foot. J Pros Ortho 3(4):150–161
7. Postema K, Hermens H, De-Vries J, Koopman H, Eisma W (1997) Energy storage and release of prosthetic feet Part 1: biomechanical analysis related to user benefits. Pros Ortho Int 21(1):17–27
8. Postema K, Hermens H, De-Vries J, Koopman H, Eisma W (1997) Energy storage and release of prosthetic feet Part 2: subjective ratings of 2 energy storing and 2 conventional feet, user choice of foot and deciding factor. Pros Ortho Int 21(1):28–34
9. Linden M, Solomonidis S, Spence W, Li N, Paul J (1999) A methodology for studying the effects of various types of prosthetic feet on the biomechanics of trans-femoral amputee gait. J Biomech 32(9):877–889
10. Hafner BJ, Sanders JE, Czerniecki J, Fergason J (2002) Energy storage and return prostheses: does patient perception correlate with biomechanical analysis? Clin Biomech 17(5):325–344
11. Hafner BJ, Sanders JE, Czerniecki JM, Fergason J (2002) Transtibial energy-storage-and-return prosthetic devices: a review of energy concepts and a proposed nomenclature. J Rehabil Res Dev 39(1):1
12. Fey NP, Klute GK, Neptune RR (2011) The influence of energy storage and return foot stiffness on walking mechanics and muscle activity in below-knee amputees. Clin Biomech 26(10):1025–1032
13. Zmitrewicz RJ, Neptune RR, Sasaki K (2007) Mechanical energetic contributions from individual muscles and elastic prosthetic feet during symmetric unilateral transtibial amputee walking: a theoretical study. J Biomech 40(8):1824–1831
14. Bedaiwi BA, Chiad JS (2012) Vibration analysis and measurement in the below knee prosthetic limb: part I-experimental work. Proc ASME Int Mech Eng Congr Exposition, 851–858
15. Cherelle P, Grosu V, Matthys A, Vanderborght B, Lefeber D (2014) Design and validation of the ankle mimicking prosthetic (AMP-) foot 2.0. Trans Neural Syst Rehabil Eng 22(1):138–148
16. Cherelle P, Grosu V, Van-Damme M, Vanderborght B, Lefeber D (2013) Use of compliant actuators in prosthetic feet and the design of the AMP-foot 2.0. Model Simul Optim Bipedal Walking, 17–30
17. Cherelle P, Junius K, Grosu V, Cuypers H, Vanderborght B, Lefeber D (2014) The amp-foot 2.1: actuator design, control and experiments with an amputee. Robo 32(8):1347–1361

18. Cherelle P, Mathijssen G, Wang Q, Vanderborght B, Lefeber D (2014) Advances in propulsive bionic feet and their actuation principles. Adv Mech Eng. https://doi.org/10.1155/2014/984046
19. Starker F, Schneider U, Hansen AH, Childress DS, Pauli J, Pauli C (2015) Artificial ankle, artificial foot and artificial leg. Google Patents
20. Ko CY, Kim SB, Kim JK, Chang Y, Cho H, Kim S, Ryu J, Mun M (2016) Biomechanical features of level walking by transtibial amputees wearing prosthetic feet with and without adaptive ankles. J Mech Sci Technol 30(6):2907–2914
21. Ghaith FA, Khan FA (2012) Nonlinear finite element modeling of prosthetic lower limbs. Proc Int Conf Adv Robo Mech Eng Des. 02.ARMED.2012.2.3
22. Jimenez-Fabian R, Flynn L, Geeroms J, Vitiello N, Vanderborght B, Lefeber D (2015) Sliding-bar MACCEPA for a powered ankle prosthesis. J Mech Rob 7(4):041011
23. Kerkum YL, Al Buizer, Noort JC, Becher JG, Harlaar J, Brehm MA (2015) The effects of varying ankle foot orthosis stiffness on gait in children with spastic cerebral palsy who walk with excessive knee flexion. PLoS ONE 10(11):e0142878
24. Leardini A, O'Connor JJ, Giannini S (2014) Biomechanics of the natural, arthritic, and replaced human ankle joint. J Foot Ankle Res 7(1):8
25. Noroozi S, Rahman AGA, Dupac M, Vinney JE (2012) Dynamic characteristics of prosthetic feet: a comparison between modal parameters of walking, running and sprinting foot. Adv Mech Design, 339–344
26. Omasta M, Palousek D, Navrat T, Rosicky J (2012) Finite element analysis for the evaluation of the structural behaviour of a prosthesis for trans-tibial amputees. Med Eng Phys 34(1):38–45
27. Rigney SM, Simmons A, Kark L (2015) Concurrent multibody and Finite Element analysis of the lower-limb during amputee running. Proc Eng Med Biol Soc, 2434–2437
28. Veneva I, Vanderborght B, Lefeber D, Cherelle P (2013) Propulsion system with pneumatic artificial muscles for powering ankle-foot orthosis. J Theor Appl Mech 43(4):3–16
29. Casillas JM, Dulieu V, Cohen M, Marcer I, Didier JP (1995) Bioenergetic comparison of a new energy-storing foot and SACH foot in traumatic below-knee vascular amputations. Arch Phys Med Rehabil 76:39–44
30. Rao SS, Boyd LA, Mulroy SJ, Bontrager EL, Gronley J, Perry J (1998) Segment velocities in normal and transtibial amputees: prosthetic design implications. Trans Rehabil Eng 6:219–226
31. Torburn L, Perry J, Ayyappa E, Shanfield SL (1990) Below-knee amputee gait with dynamic elastic response prosthetic feet: a pilot study. J Rehabil Res Dev 27:369
32. Lehmann J, Price R, Boswell-Bessette S, Dralle A, Questad K, DeLateur B (1993) Comprehensive analysis of energy storing prosthetic feet: flex foot and seattle foot versus standard SACH foot. Arch Phys Med Rehabil 74:1225–1231
33. Macfarlane PA, Nielsen DA, Shurr DG, Meier K (1991) Gait comparisons for below-knee amputees using a flex-foot versus a conventional prosthetic foot. J Pros Ortho 3:150–161
34. Au S, Berniker M, Herr H (2008) Powered ankle-foot prosthesis to assist level-ground and stair-descent gaits. Neural Netw 21:654–666
35. Sun J, Voglewede PA (2014) Powered transtibial prosthetic device control system design, implementation, and bench testing. J Med Devices 8:011004
36. Collins SH, Kuo AD (2010) Recycling energy to restore impaired ankle function during human walking. PLoS ONE 5:e9307
37. Howell LL, Magleby SP, Olsen BM (2013) Handbook of compliant mechanisms. Wiley
38. Dao TP, Huang SC (2015) Design, fabrication, and predictive model of a 1-dof translational, flexible bearing for high precision mechanism. Trans Canad Soc Mech Eng 39(3):419–429
39. Dao TP, Huang SC (2017) Compliant thin-walled joint based on zygoptera nonlinear geometry. J Mech Sci Technol 31(3):1293–1303
40. Dao TP (2016) Multiresponse optimization of a compliant guiding mechanism using hybrid Taguchi-grey based fuzzy logic approach. Math Prob Eng. https://doi.org/10.1155/2016/5386893
41. Huang SC, Dao TP (2016) Design and computational optimization of a flexure-based XY positioning platform using FEA-based response surface methodology. Int J Precis Eng Manuf 17(8):1035–1048

42. Dao TP, Huang SC (2016) Design and analysis of a compliant micro-positioning platform with embedded strain gauges and viscoelastic damper. Microsyst Technol, 1–16
43. Huang SC, Dao TP (2016) Multi-objective optimal design of a 2-DOF flexure-based mechanism using hybrid approach of grey-taguchi coupled response surface methodology and entropy measurement. Arabian J Sci Eng 41(12):5215–5231
44. Dhote S, Jean Z, Yang Z (2015) A nonlinear multi-mode wideband piezoelectric vibration-based energy harvester using compliant orthoplanar spring. Appl Phys Lett 106(16):163903
45. Dhote S, Zhengbao Y, Jean Z (2018) Modeling and experimental parametric study of a tri-leg compliant orthoplanar spring based multi-mode piezoelectric energy harvester. Mech Syst Sig Process 98:268–280
46. Bains PS, Singh S, Sidhu SS, Kaur S, Ablyaz TR (2018) Investigation of surface properties of Al–SiC composites in hybrid electrical discharge machining. In: Futuristic composites. Springer, Berlin, pp 181–196
47. Bhui AS, Singh G, Sidhu SS, Bains PS (2018) Experimental investigation of optimal ED machining parameters for Ti-6Al-4 V biomaterial. FU Ser Mech Eng 16(3):337–345
48. Sidhu SS, Bains PS, Yazdani M, Zolfaniab SH (2018) Application of MCDM techniques on nonconventional machining of composites. In: Futuristic composites. Springer, Berlin, pp 127–144
49. Nguyen TT, Dao TP, Huang SC (2017) Biomechanical design of a novel six dof compliant prosthetic ankle-foot 2.0 for rehabilitation of amputee. ASME Int Des Eng Tech Conf Comp Info Eng, V05AT08A013–V05AT08A013
50. Nguyen TT, Le HG, Dao TP, Huang SC (2017) Evaluation of structural behaviour of a novel compliant prosthetic ankle-foot. IEEE Int Conf Mech Sys Cont Eng, 58–62

# Chapter 7
# Biomaterials in Tooth Tissue Engineering

C. Pushpalatha, Shruthi Nagaraja, S. V. Sowmya and C. Kamala

## 1 Introduction

Tooth is a biological organ originating from ectomesenchymal cells composed of enamel, dentin, and viable pulp tissue which is altogether called as tooth organ. These tissues usually arise from the interaction of oral epithelium and mesenchyme of cranial neural crest. Ameloblasts of ectodermal origin generally deposit enamel and disappears once hard tissue formation completes. Odontoblast cells of ectomesenchymal origin secrete dentin and persist as dental pulp for the whole life. Tooth loss and dental diseases resultant of caries, trauma, periodontal disease, and genetically inherited disease are considered as a major health issues. The traditional approach of dental treatment uses non-biological substitutes which build up the structure, function, and esthetics component of the tooth but do not promote biological vitality in the tissues. The resultant tooth will be non-vital which is devoid of physiological defense mechanisms and nerves for pain transmission. In the case of immature teeth, the

C. Pushpalatha (✉)
Department of Pedodontics and Preventive Dentistry, Faculty of Dental Sciences,
M.S. Ramaiah University of Applied Sciences, Bengaluru, India
e-mail: drpushpalatha29@gmail.com

S. Nagaraja
Department of Conservative Dentistry and Endodontics, Faculty of Dental Sciences,
M.S. Ramaiah University of Applied Sciences, Bengaluru, India
e-mail: shruthi.cd.ds@msruas.ac.in

S. V. Sowmya
Department of Oral Pathology and Microbiology, Faculty of Dental Sciences,
M.S. Ramaiah University of Applied Sciences, Bengaluru, India
e-mail: drsowmya25@gmail.com

C. Kamala
Department of Medical Electronic, Dr. Ambedkar Institute of Technology, Bengaluru, India
e-mail: kamala_ray@yahoo.co.in

© Springer Nature Singapore Pte Ltd. 2019                                      91
P. S. Bains et al. (eds.), *Biomaterials in Orthopaedics and Bone Regeneration*,
Materials Horizons: From Nature to Nanomaterials,
https://doi.org/10.1007/978-981-13-9977-0_7

ability to undergo further root formation is hampered. These issues focused toward the development of new biological approach is of concern in vital tissue regeneration. The tissue engineering has focused significantly on the replacement and regeneration of entire tooth or individual components of a tooth which is structurally and functionally sound. The molecular signals associated with odontogenesis and tooth-related phenotypes lead to concept of tooth tissue engineering. Tooth regeneration through advanced and innovative approaches eliminates the pitfall of treatment outcomes linked with traditional approaches. Significant progress toward successful tooth regeneration has been achieved by utilizing a combination of tissue-specific stem cells or progenitor cells, signaling molecules, and biomaterial scaffolds that are the guiding principles in tooth tissue engineering.

## 2    Strategies for Tooth Regeneration

Forsyth Group in teamwork with Dr. Vacanti in 2002 first achieved tooth regeneration through classical tissue engineering techniques. The dental tissue engineering approaches generate bioengineered replacement teeth and dental tissues mimicking natural tooth by understanding and applying the regulatory mechanism of tooth developmental processes. The formed functional bioengineered tooth has the capability to occlude accurately in dentition with potential interproximal contacts, proprioceptive function, efficient transmission of masticatory forces, and esthetics reestablishment. To regenerate bioengineered teeth with predetermined morphology, it is mandatory for systematic arrangement of epithelial–mesenchymal cell layers onto the scaffold and its interaction with the extracellular matrix. Three variants of dental tissues are essential for tooth regeneration such as dental epithelium, dental pulp, and dental follicle which are associated with enamel, dentin, pulp, and periodontium formation. A range of advanced tooth tissue engineering techniques can be accomplished using these cells through its implantation and interaction with the extracellular matrix. Tooth regeneration techniques involving regulatory processes of odontogenesis are scaffold-based and scaffold-free approaches.

## 2.1    Scaffold-Based Approach

In this approach, the cells are seeded in an orderly manner on the scaffold in the laboratory. The seeded scaffolds are precultured in vitro for few days and transplanted in vivo into animal tissues such as renal capsule or omentum to obtain adequate blood supply, oxygen, or nutrients (Fig. 1). To promote diffusion of metabolites and its nutrients, the cells can be cultured under perfusion or flow perfusion bioreactors. The desired recipient site can later be grafted with regenerated tissue or organ to obtain an organ of specific shape and size to fulfill the mechanistic requirement at the recipient site. The construct is implanted into the jaw and designed to overcome the

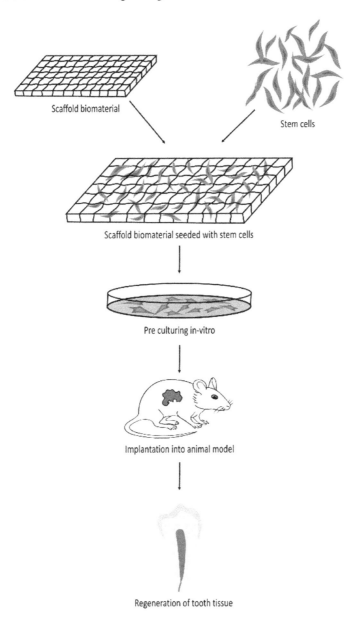

Scaffold biomaterial

Stem cells

Scaffold biomaterial seeded with stem cells

Pre culturing in-vitro

Implantation into animal model

Regeneration of tooth tissue

**Fig. 1**  Scaffold-based approach for tooth regeneration

mechanical and physiological stresses. This approach results in three-dimensional scaffold that is rich in growth factors and aids in cell adhesion, proliferation, differentiation, and migration thereby imparting mechanical support for extracellular matrix tissue regeneration [1]. Young et al. [2] first successfully bioengineered tooth crowns by seeding dissociated cells on poly (L-lactide-co-glycolide) (PLGA) biodegradable polymer scaffolds. The histological, immunohistochemical, and molecular analyses have shown that the bioengineered tooth contained morphologically similar enamel organ, dentin, predentin, odontoblasts, pulp chamber, and Hertwig's epithelial root sheath with proper cellular and protein organization as observed in natural tooth. Duailibi et al. [3] suggested that tooth tissue engineering methods can regenerate tooth tissues in both pig and rat by culturing tooth bud cells of postnatal rat and by seeding them on biodegradable scaffold such as PLGA or polyglycolic acid (PGA). It was also demonstrated that during tooth morphogenesis, the tooth cusp number could be controlled by cap stage dental mesenchyme and reaggregation of the mesenchymal and dental epithelium. However, the shortcoming of these studies was that the regenerated tooth did not have morphological size and shape mimicking natural tooth. Honda et al. [4] employed a novel technique wherein epithelial and mesenchymal cells were seeded sequentially such that there is a direct contact between the epithelium and mesenchyme. It was proposed that the cell seeding technique can be beneficial in regulating the tooth morphology. Duailibi et al. [5] cultured cells from tooth bud and seeded them on tooth-like molds composed of PGA/PLGA/poly-L-lactide (PLLA) in the jaw. Subsequently 12 weeks after, an organized tooth crown similar to that of implants grown in the omentum was formed. The resultant tooth displayed physical properties and functions similar to natural teeth. Seeding of dental bud cells on gelatin chondroitin hyaluronan tri-copolymer scaffold regenerated tooth structures with a set of proteins necessary for odontogenesis. Although several studies show that the bioengineered teeth with organized dental tissue in the seeded scaffold are effective, the main disadvantage is that the regenerated tooth is repeatedly very small and occasionally simulates the natural tooth in morphology. The translation of this cell transplantation technique into therapeutics is quite challenging due to the following reasons.

- Autologous tooth germ cells are inaccessible for humans.
- Xenogenic tooth germ cells may induce immune rejection and dental abnormalities.
- Limited availability of autologous postnatal tooth germ cells or dental pulp stem cells and translational barriers.
- Difficulties in regulatory approval and excessive commercialization cost.

To overcome the limitations of cell-based therapy, an alternative approach without the use of cell delivery system called stem cell homing was first introduced in 2010 to regenerate anatomically shaped tooth-like structures [6]. This approach utilizes the in vivo induction of mesenchymal stem cells with specific target localization which will be circulating in the blood and differentiate into conforming tissues. This therapeutic approach does not involve isolation and ex vivo manipulation of cells in the laboratory, thereby enhancing clinical, commercialization, and regulatory processes.

Cell homing bring about tissue regeneration in the injured site by chemotaxis through biomolecular signals. Kim et al. [7] was the first to show that chemotaxis-induced cell homing is suitable for pulp tissue regeneration in an endodontically treated tooth. This method is commonly used in dental pulp and periodontal ligament regeneration and forms a noticeable clinical translational approach. In pulp regeneration, growth factors impregnated in bioactive scaffolds are injected into root canals of the teeth to induce cell migration, proliferation, and differentiation of stem cells [8–10]. Growth factors like stromal-derived factor-1 (SDF1) and bone morphogenetic protein (BMP)-7 are delivered into the scaffold to recruit endogenous cells and promote angiogenesis for periodontal ligament regeneration [11].

## 2.2 Scaffold-Free Approach

This approach involves encouraging embryonic tooth formation processes with appropriate signaling molecules to produce tooth tissue that resembles natural tooth anatomically. In 2002, Ohazama et al. [12] developed artificial embryonic tooth primordia after transplantation into adult jaw using recombinants such as non-dental-derived mesenchymal cell and oral embryonic epithelium. They also showed that non-dental stem cells such as embryonic stem cells, neural stem cells, and bone marrow-derived stem cells can develop tooth structure by the expression of odontogenic genes. They also were the first to develop bioengineered tooth that mimicked natural tooth histologically by the implantation of E14.5 rat tooth primordia into adult mice. Various studies report that epithelial–mesenchymal interactions are important parameters for tooth development [13, 14]. In scaffold-free studies, interaction of epithelial–mesenchymal cells is permitted and tissue is developed through cell aggregation without any external carrier material for cellular attachment [15, 16]. The main limitation associated with this approach is that large quantities of mesenchymal cells are essential to develop morphologically distinct tooth structures. An improved bioengineered three-dimensional organ-germ culture method was developed to generate ectodermal organs such as teeth and whisker follicles by reciprocal interactions of epithelial and mesenchymal cells. After transplantation of bioengineered tooth germ into subrenal capsules in mice, the tooth primordium generated plural incisors composed of enamel, ameloblasts, odontoblasts, dentin, dentinal tubules, pulp tissue, Tomes' processes, root, periodontal ligaments, alveolar bone, and blood vessels which were organized in similar way to natural tooth on histological observation. The bioengineered tooth germs show an organized arrangement of mineralized tissues, cell types, detectable blood vessels, and nerve fibers in the pulp. This approach produces oral epithelium-like structures or bone but not the whole tooth. Ikeda et al. [17] successfully demonstrated the development of fully functional bioengineered tooth through regenerative therapy using 3D organ culture method. The developed tooth had a similar hardness to the adult natural tooth with normal colony-stimulating factor-1 and parathormone-1 gene expression which could regulate osteoclastogenesis. The nerve fibers such as anti-neurofilament,

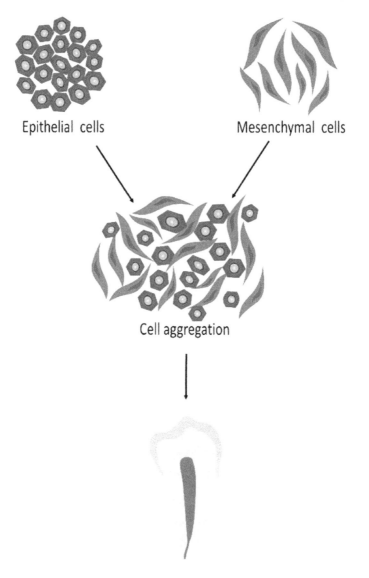

Fig. 2 Scaffold-free approach for tooth regeneration

neuropeptide Y, calcitonin-gene-related peptide, and galanin-immunoreactive neurons effectively re-entered the pulp and periodontal regions which responded to noxious stimuli such as pain and mechanical stress. This study also provided an insight into the tooth eruption mechanisms and masticatory potential of a bioengineered tooth. Oshima et al. [18] developed a bioengineered mature tooth unit possessing physiological tooth functions in the subrenal capsule of the mouse using a size-control device (Fig. 2).

# 3  Innovative Approaches for Enamel Regeneration

In addition to scaffold-based and scaffold-free approaches, enamel tissue regeneration includes cell pellets culture system, recombination experiments, self-assembly bioactive nanomaterials, and gene-based enamel formation [19, 20].

## 3.1  Cell Pellet Culture System

Pellet culture system is a beneficial model developed to trigger dental papilla mesenchymal cells (DPMCs) at the advanced bell stage to differentiate dentin–pulp complexes with enamel-shaped cover to promote dentinogenesis on seeding into the renal capsule [21]. Similar study by Iohara et al. [22] showed that cell pellet culture system can provide a three-dimensional growth condition which is essential for cell differentiation and produce a better yield of extracellular matrix which acts as natural scaffolds in cell pellets. Matrigel composed of laminin which forms the principal part of pellet culture system has been used as a substitute to biodegradable scaffolds [23, 24]. Although Matrigel by itself does not promote biomineralization and generate enamel tissue, it can accelerate the expression of adhesion molecules and receptors for ameloblast lineage cells. Human embryonic stem cells along with BMP-4 and retinoic acid, embedded in the Matrigel-coated culture will differentiate into epithelium-like cells, without demonstrable amelogenin. However, ameloblast differentiation demands more complex signals for increased expression of amelogenin, integrin $\alpha2$, and integrin $\beta7$ with the absence of biomineralized enamel [25, 26].

## 3.2  Recombination Experiments

Recombination experiments were carried out by simulating epithelial–mesenchymal cell interactions during morphogenesis using reciprocal signaling networks. Recombination of epithelial and mesenchymal tissues from different animal species has been attempted to form enamel suggesting that odontogenesis takes place on reactivation of genes by the mesenchyme of different species [27]. Recombination of epithelial cells and mesenchymal tissue results in tooth-like morphology displaying normal cusp number and crown shape [28]. Similarly, epithelium layers obtained from mandible primordia of E10 mice induced the development of dental tissue including enamel with proper organization of a tooth. Recombination of mesenchymal and epithelial cells has been effective in developing ameloblasts that could form well-organized enamel in a bioengineered tooth. Epithelial and mesenchymal cells dissociated from E14.5 mice seeded within a collagen gel drop and then transplanted into subrenal capsules or cultured in vitro led to the formation of ameloblasts which secreted organized enamel in a bioengineered tooth.

## 3.3 Artificial Bioactive Nanomaterials

Self-assembly peptide amphiphiles (PAs) molecules carrying Arg-Gly-Asp (RGD) peptide were developed for cellular receptor signaling. To induce biological adhesion, the fibronectin RGD epitope is commonly displayed [29, 30]. Huang et al. [31] showed the interaction between ameloblasts or their progenitor cells and PAs through RGD peptide signals. Peptide motif RGD and PAs molecules self-assembled together into nanofiber-induced biomineralization by the ameloblast-like cell line. It showed more mineral formation through increased expression of integrin α5 mRNA and amelogenin upregulation.

## 3.4 Gene-Manipulated Enamel Regeneration

Borovjagin et al. [32] reported that gene-manipulated treatment was effective in the treatment of amelogenesis imperfecta using vectors and modified delivery system. The localized gene manipulation can re-establish the mineralization procedures by temporally targeting important components in enamel development. It is proven that gene therapy might be a feasible approach for enamel regeneration or repair especially during permanent tooth development. In another study, oral epithelial and odontoblast mesenchymal cells were reprogrammed using Pitx2 and miR-200a-3p into a single type of cell for enamel regeneration. This preprogrammed approach leads to amelogenin expression, Sox2 upregulation, and decreased mesenchymal markers expression, but still regeneration of enamel has not been reported [33]. The study reports that thymosin beta-4 was as an active gene during tooth germ development and TMSB4X-transfected HaCaT cells have mineralizing ability [34, 35]. Several studies have employed gene-manipulated treatment using Ad5-pK7/RGD, Pitx2, miR-200a-3p, TMSB4X as vectors and part of the modified delivery system to regenerate and repair enamel during tooth development. However, successful results could not be achieved as the enamel protein expression was not concordant with that of normal enamel secretion [36].

# 4 Innovative Approaches for Dentin Regeneration

The inductive molecular signals associated with epithelium are essential for formation of dentin by inducing odontoblasts to produce dentin matrix proteins. Initially, demineralized bone powder was placed on exposed pulp surface to induce mineralization, but was found to be ineffective [37]. BMP present in the dentin is required for stimulation of odontoblasts for formation of reparative dentin. It was found that area of reparative dentin induced is proportionally dependent on the quantity of BMP used, which predetermines the dentin mass formed. Dentin regeneration can be induced using human recombinant proteins with collagen matrices, whereas in case of inflamed pulp, the reparative dentin formation is not effective since active recombinant protein possess short half-life and rapid protein degradation. Growth factor 11 can also induce reparative dentin formation by placing them directly onto the pulp cells. Additionally, bone sialoprotein aids information of reparative dentin through laying down of extracellular matrix by the differentiating dental pulp cells. The reparative dentin laid down by bone sialoprotein exhibits varied morphological features. Isolated stem cells from pulp and periodontal ligament grafted into defects show partial regeneration of dentin and periodontal-like tissue when appropriate signals coexist [38, 39].

Dentin–enamel regeneration using scaffold and cell aggregate methods employs dissociated cells. Shinmura et al. [40] showed that epithelial cell rests of Malassez (ERM) have the capacity to regenerate dental tissues like enamel, stellate reticulum, and ameloblast-like cells by transplanting cultured ERM seeded onto collagen sponge scaffolds into the rat omentum. The complex dental tissue regeneration was examined by reassociation between human dental epithelial stem cells and human dental pulp stem cells from mouse embryos which were cultured in vitro initially and later implanted in vivo.

# 5 Innovative Approaches for Dentin–Pulp Complex Regeneration

Pulpal regeneration includes three crucial steps: (1) Destructed pulpal tissues must carefully be disinfected, and all microorganisms must be eradicated through antimicrobial therapy; (2) inflammation control at different levels of the tissue injury; (3) successful regeneration of lost pulpal tissue. In the last step, the endodontic regenerative biomaterials should encourage cell migration and proliferation of the different components at the tissue injury [41–43]. Pulpal regeneration strategies rely upon the state of the dental pulp, the stage of inflammation, and extent of tissue infected and damaged. In initial lesion with pulp exposure leading to dentin loss and less destruction of pulp, regeneration is performed using pulp capping biomaterials such as calcium hydroxide, mineral trioxide aggregates, and biodentin. These biomaterials encourage stem cells of the pulp to differentiate into odontoblast-like cells or to secrete transforming growth factor (TGF)-b131.

The regenerative strategy can be applied successfully to induce apical closure (apexification) in immature open apex with necrotic pulp tissue by placing biomaterials like calcium hydroxide and mineral trioxide aggregate in the coronal or in the deeper portion of the pulp [44]. The same regenerative biomaterials can be used to encourage physiological root development (apexogenesis) in the case of vital pulp tissue. In the case of a significant lesion with pulpal infection and inflammation, the revascularization method is employed to promote pulpal vitality. In immature teeth with open apex, bleeding is induced that results in filling of the root canal with a blood clot that acts as a scaffold for pulp and promotes root formation. The major pitfall of this approach is that the regenerated pulp tissue is different from the physiologic counterpart. To overcome this drawback, stem cell therapy is advocated for dentin–pulp complex regeneration since dental tissues are abundant with stem cells. To carry out pulp regeneration, different types of stem cells are used such as postnatal dental pulp stem cells (DPSCs), stem cells from exfoliated deciduous teeth (SHED), stem cells from apical papilla (SCAP), periodontal ligament stem cells (PDLSCs), and dental follicle progenitor cells. As DPSCs, SHED, and SCAP are derived from pulp or precursor of pulp, these stem cells are more suitable for dentin–pulp regeneration.

Stem cell-sheet-derived pellet and encapsulated form of stem cells have been tried for dentin–pulp complex regeneration. The commonly tried cell encapsulated materials are enzyme-cleavable, customized self-assembled peptide hydrogels, glycolide, polyethylene gylated fibrin hydrogels, or biodegradable lactide. Gelfoam-encapsulated dental stem cells have been found to be effective in forming dentin–pulp complex in case of non-vital root canals especially in young permanent incisors of beagles. The dentin–pulp complex can be stimulated using Emdogain and platelet-rich plasma combination. Scaffolds encompassed by growth factors such as endothelial growth factor (EGF), basic fibroblast growth factor (bFGF), and TGF-b1 enhance the stem cells function. Bioactive factors can be accumulated or encapsulated and allowed to release from biomaterials in vivo. The adjacent and or systemic cells present next to the root apices of root canal treated teeth migrate into the anatomic part and allow neovascularization and reinnervation. In this case, the root canal of the tooth acts as a scaffold. Single or multiple cytokines, platelet-derived growth factor (PDGF), vascular endothelial growth factor (VEGF), or bFGF, alone or in combination with nerve growth factor (NGF), and BMP-7 play a role as critical signaling molecules in pulp regeneration. It is found that the implantation on calvarial site is superior to the dorsum for pulp regeneration using cell homing approach. Injectable scaffolds are excellent candidates for pulp and dentin regeneration since the tooth root canals are small and of irregular shape. Collagen, fibrin, synthetic self-assembled nanofibrous peptide hydrogel (PuraMatrix™), self-assembling peptide amphiphiles, RGD-modified PAs, and polyethylene glycol maleate citrate hydrogel were used as an injectable scaffold for dentin–pulp complex regeneration. As a first step, a novel anatomically shaped human mandibular first molar scaffold was fabricated with communicating microchannels, 3D layer-by-layer apposition for angiogenesis, and homing of host cells. Suzuki et al. [45] achieved regeneration of dental pulp tissue by recruiting and subsequently differentiating by chemotaxis of selective cytokines such as SDF1, bFGF, or BMP-7. Pan et al. [46] first documented the expression of stem cell factor (SCF) and c-Kit receptor CD117 in dental pulp cells, in the dental follicle, and in periodontal ligament progenitors. It was also observed that application of SCF in regeneration induced chemotaxis, angiogenesis, tissue remodeling, and synthesis of collagen matrix. Takeuchi et al. [47] showed that bFGF is potentially useful in pulp regeneration as cell homing or migration factor is similar to the influence of granulocyte colony-stimulating factor. Yang et al. [48] provided novel in vivo ectopic transplantation model for pulp regeneration and showed that SDF-1a-loaded silk fibroin scaffolds have a promising effect on pulp revascularization by enhancing DPSCs migration, focal adhesion, and stress fiber assembly through autophagy. Zhang et al. [49] observed that cell homing of systemic bone marrow stromal cells to the root canal surface contributed to dental pulp-like tissue regeneration. It was also analyzed that SDF-1 on intracanal application enhanced the efficiency of bone marrow stromal cells homing and angiogenesis. It was also found that stem cell factor in human immature teeth can accelerate cell homing and maturation of the dentin–pulp complex. Li and Wang [50], in an in vivo study, used a combined approach of PDGF-BB, nerve growth factor, and a brain-derived neurotrophic factor for successful dental pulp-like tissue regeneration ectopically by cell homing without using any exogenous cells (Fig. 3).

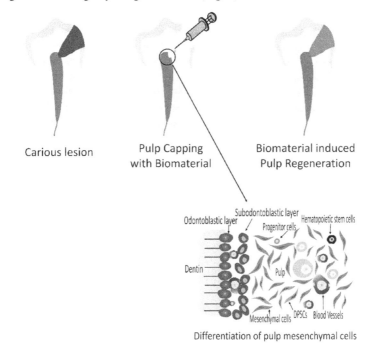

**Fig. 3**  Pulp regeneration using biomaterial

# 6  Innovative Approaches for Periodontal Regeneration

In early 1980s to restore the lost periodontal tissues, demineralization of cementum of the tooth was carried out to expose inserted collagen fibers for the integration of new and exposed collagen fibers. The main disadvantage associated with this therapy is ankylosis and root resorption [51]. Hence, utilization of bone grafts into the periodontal defects was adopted, but these materials led to minimal osteoinductive capacity enveloped by dense connective tissue of fibrous nature. Later guided tissue regeneration procedure was employed by draping the biomaterial membrane into the periodontal bony defect extending from the root to the adjacent bone surface. This procedure allows regeneration of the bony defect by migration of cells located in periodontal ligament. Researchers have improved this therapy by incorporating growth factors and stem cells. The common growth factors used for regeneration of periodontal tissues are BMPs, Emdogain, PDGFs, and recombinant amelogenin protein. Further successful demonstration of root periodontal tissue was achieved using pelleted hydroxyapatite or tricalcium phosphate scaffold composed of dental stem cells of the apical papilla. Transplanted cells seeded on multilayered polyglycolic acid sheets when inserted into root surfaces generated new bone, root cementum, and well-organized collagen fibers. In addition to periodontal ligament-derived dental stem cells, other cells such as bone marrow-derived mesenchymal stem cells and adipose-derived stem cells are shown to encourage periodontal regeneration. The combination of transplanted mesenchymal stem cells and exogenous molecular signals is found to be more effective in periodontal tissue regeneration.

# 7  Challenges in Tooth Regeneration

The three main challenges associated with tooth regeneration are difficult accessibility of clinically applicable cell source, tough to induce odontogenic capacity in these cells, and accelerate total developmental process by gene control. Sufficient quantities of cell sources are essential for tooth regeneration especially DPSCs, SHED, SCAP, and PDLSCs. These stem cells are deficient since dental cells of epithelial origin go through apoptosis subsequently to enamel formation completion and dental epithelial expansion is a difficult task in ex vivo. Hence, non-dental cells were identified for regeneration. It is hard to obtain an applicable embryonic stem (ES) cells source in particular to the patient along with ethical concern. The xenogenically derived ES cells commonly show immune rejection upon clinical use. Autologous-based postnatal stem cells from embryonic tooth germ are difficult to isolate and expand which may not be clinically useful. In addition, these cells lose odontogenic potency after in vitro expansion and may not act as an appropriate source for regeneration of a precise tooth structure. The discovery of induced pluripotential stem cells (iPSc) eliminated ethical concerns associated with ES cells and limited adult stem cells pluripotency. However, the use of iPSC in host tissue has shown cancer-like growth. The molecular signaling networks have been tried for tooth development through epithelial mesenchyme interactions but are minimally understood. Hence, clinically tooth regeneration is not possible via stimulating signaling molecules.

# 8 Stem Cells in Dental Tissues

The existence of stem cells within the tooth has been reported for the first time in 2000 [52]. Epithelial stem cell niche in teeth was first isolated from apical end of rodent incisor via organ culture. In humans, the stem cells include adult epithelial and mesenchymal cells, which are essential for regeneration of human tooth. Shi et al. isolated stem cells from adult dental pulp with high self-renewal and multiple cell lineages potential. Dental mesenchymal stem cells and bone marrow-derived mesenchymal stem cells are the two varieties of human adult mesenchymal stem cells used for tooth regeneration. Both these cells have osteo/odontogenic cell lineage capacity analyzed during the process of tooth engineering. Among them, dental mesenchymal cells have been recognized for tooth regeneration because of more odontogenic than osteogenic potential [53]. The dental mesenchymal stem cells in tooth regeneration are stem cells from the DPSCs, SCAP, PDLSCs, SHED, and dental follicle precursor cells (DFPCs). The other non-dental stem cell sources available in oral cavity and used in regenerative dentistry are oral mucosa-derived stem cells, salivary gland-derived stem cells, alveolar bone-derived mesenchymal stem cells, and adipose tissue-derived stem cells with minimal available literature evidence.

## 8.1  Dental-Derived Stem Cells

### 8.1.1  Stem Cells from Dental Pulp

Tooth-related DPSCs were isolated first from the pulp of permanent third molars in 2000 by Gronthos et al. [54]. These isolated cells were able to differentiate into various derivates of mesenchymal cells like odontoblasts, adipocytes, chondrocytes, and osteoblasts [55]. DPSCs can be isolated from normal and inflamed pulps with differences in cellular characteristics. DPSCs can also be obtained from aged pulp tissues, and its potency can be maintained using growth factors and nanostructured hydroxylapatite scaffolds [56]. DPSCs exhibit favorable odontogenic potential and generate tissues with morphological and functional features resembling human dental pulp. Literature evidence supports the development of a three-dimensional model representing early odontogenesis through human normal oral epithelial cells and DPSCs [57]. DPSCs are considered to be ideal for tissue reconstruction since it can be obtained with greater efficiency, considerable differentiation capacity, and demonstrable interaction with biomaterials. Scaffold-free three-dimensional cell constructs containing DPSCs reveal a vital part in dental pulp regeneration with increased vascularity.

### 8.1.2 Apical Papilla-Derived Stem Cells (Stem Cells from Apical Papilla)

Stem cells from the apical papilla (SCAPs) were first reported by Abe et al. in the apical papilla of the roots of developing teeth. The third molars and teeth with open apices are an important source of SCAPs. These cells have high proliferation and multilineage differentiation activity in vitro in comparison to DPSCs. These cells have the ability to differentiate into other cells such as osteoblasts, odontoblasts, and adipocytes. The differentiation of APDCs into dentinogenic cells has been established using animal models. APDCs showed odontoblast-like cells differentiation which secreted dentin in vivo and presented as a primary cell source for root dentin formation [58]. The discovery of SCAP cells led to pulp tissue revascularization concept wherein apical papilla is stimulated and regeneration of damaged or lost pulp tissue through cell homing strategy was executed [59]. The SCAP holds the upper hand in dentin pulpal regeneration compared to DPSCs and SHED. SCAP brings dentin–pulp regeneration by migration of these cells onto the dentin surface and releasing growth factors embedded in dentin which acts as molecular signals for further differentiation of cells into odontoblast-like cells. The cell processes of each differentiated cell spread into the dentinal tubules, and these differentiated cells generate extracellular matrix onto the surface of dentin and into the dentinal tubular spaces. SCAP has been tested with dental bioceramics and showed favorable results toward its differentiation potential [60]. SCAP showed a favorable response to platelet concentrate, i.e., concentrated growth factor and platelet-rich fibrin and showed increased proliferation and differentiation potential [61].

### 8.1.3 Periodontal Ligament Stem Cells

PDLSCs belong to MSCs subfamily residing in the perivascular space of the periodontium. These cells were first isolated by Seo et al. [62] from human impacted third molars showing differentiation into periodontal ligaments, alveolar bone, cementum, peripheral nerves, and blood vessels. It is found that orthodontic forces can act as a positive modulator and increase the proliferation of PDLSCs and expression of osteogenic and angiogenic factors [63]. It has also been observed that dental bioceramics improve the osteogenic potential of PDLSCs [64]. Chitosan films have shown to promote higher self-renewal, gene expression, osteogenic capacity, and colony-forming units of PDLSCs [65].

### 8.1.4 Dental Follicle Progenitor Cells (DFPCs)

Dental follicle cells are ectomesenchymal cells surrounding developing tooth germ which can be easily isolated from an impacted third molar. The presence of stem cells in the dental follicle was first reported in 2002 [66]. It is believed that dental follicle tissue contains progenitor cells for cementoblasts, periodontal ligament cells, and osteoblasts. DFPCs have been reported to exhibit better proliferation, colony-forming ability, and differentiation compared to SHED and DPSCs [67].

### 8.1.5 Stem Cells from Exfoliated Deciduous Teeth (SHED)

Batouli et al. [68] isolated a distinctive population of multipotent stem cells from the pulp remnants of exfoliated deciduous teeth. These cells have the capacity to promote bone formation and secretion of dentin and possess the ability to differentiate into non-dental mesenchymal cells. Immature dental pulp stem cells (IDPSCs) are kind of SHED cells which are isolated from primary teeth. SHED exhibits higher proliferation rates compared to DPSCs and bone marrow-derived MSCs. SHED has different odontogenic and osteogenic differentiation potential than DPSCs. Although SHED cells are incapable of osteoblastic or osteocytic differentiation, they induce osteogenic differentiation thereby demonstrating its osteoinductive potential. Therefore, SHED can be used in vivo in dental pulp tissue engineering, where stem cells replace the infected pulp, resulting in architecturally and cellularly resembling normal dental pulp [69]. SHED shows high expression for genes involved in cell proliferation and extracellular matrix formation like TGF-β and fibroblast growth factor [70]. The exfoliated teeth are one of the sources for SHED cells that can be used in regenerative dentistry by commercial banking of these stem cells. The preserved SHED cells from exfoliated teeth can be used once the child becomes adult for autologous and allogeneic cell replacement.

### 8.1.6 Tooth Germ Progenitor Cells (TGPCs)

These cells are novel stem cell population that was identified in the dental mesenchyme of the third molar tooth germ during the advanced bell stage. TGPCs have adipogenic, chondrogenic, osteogenic/odontogenic, neurogenic differentiation capacity and able to form tube-like structures, possibly an evidence of vascularization.

## 9 Selection of Biomaterials for Tooth Regeneration

Biomaterial scaffold is most crucial for tooth regeneration and should satisfy definite general requirements. To achieve significant regeneration of tooth, the biomaterial scaffolds should possess following requirements:

- Easy to handle, biocompatible, non-toxic, and good physical and mechanical stability.
- Lesser immunogenicity and should enhance vascularity.

- Biomaterial scaffold should have acceptable pore size, shape, and volume for cells and/or growth factors diffusion and passage of nutrients and waste products in the cells.
- Biologically safe biodegradability correlating with the rate of new tissue formation without any noxious by-products.
- Biomaterial scaffold should exhibit cellular encapsulation or cellular surface adhesion that regenerate single or multiple dental tissues.
- Permit functionality of at least few multiple cell types like ameloblasts, odontoblasts, fibroblasts, cementoblasts, vascular cells, and neural endings.
- As tooth organ exhibits diversity in structure and function, multiple polymeric layers may be chosen for regeneration which may be either natural, synthetic or hybrid polymers.
- Manifest clinical suitability which is readily sterilized and stored in clinical setting with reasonable shelf life.

## 10    Biomaterials for Tooth Regeneration

The biomaterials are developed to contribute an ideal platform for cell adhesion, proliferation, and differentiation. Biomaterials are basically involved in osteo/odontoconduction, osteo/odontoinduction, and osteo/odontogenesis based on their composition [71, 72]. Biomaterials can be classified into natural and synthetic based on its derivatives. Based on chemical composition, biomaterials can be classified as metals, polymers, ceramics, and composites.

### *10.1   Natural Biomaterials*

#### 10.1.1   Bioceramics

Bioceramics derived from coral or marine sponges have unique interconnected sponges which possess a significant amount of water and helps in the flow of fluid mimicking ideal bone/dentin scaffold. The sponge in the organic component of hard tissues is similar to that of vertebral column and is suitable for tissue regeneration. Biosilica, another mineral component, is able to provide cell proliferation, mineralization, and bone formation [73]. They are hard and brittle with excellent compressive strength, high resistance to wear, and low frictional properties but poor tensile strength. They possess better mechanical properties yet retain their porous structure which benefits cell adhesion, infiltration, and improved vascularization. The naturally derived ceramics have greater biocompatibility compared to synthetic. These materials help in attachment, growth, and differentiation of osteoprogenitor cells.

## 10.1.2   Natural Polymers

Proteins

Natural proteins derived from collagen, gelatin, silk fibroin, and fibrin are the types of polymers used in bone/tooth tissue engineering. Collagen and gelatin are the most commonly used natural proteins in tissue engineering because of their higher biocompatibility, lesser antigenicity, lesser cytotoxicity, and lesser inflammatory reaction. Collagen promotes cellular adhesion, cellular migration, and cell growth. Although collagen does not have high physical strength, it has high tensile strength and hence can be used for pulp regeneration. The drawback with collagen-based scaffold when used in the regeneration of pulp is high degradation rate by collagenase enzyme and noticeable contraction of collagen. To overcome this disadvantage, it is usually coupled with other polymers or cross-linked with other chemicals. Chemical cross-linking of collagen can be done with glutaraldehyde or diphenylphosphoryl azide as it not only enhances the mechanical stiffness but also compromises cell survival and biocompatibility. The mechanical properties and bone conductivity can be enhanced by hybrid scaffolds with β-tricalcium phosphate (TCP)/polyethylene and HA. The pulp with hard tissue in organized manner can be induced by combining DPSCs, collagen scaffold, and dentin matrix protein-1. The osteodent in formation can be induced by combining collagen matrix with BMP-2 or BMP-4. Collagen scaffolds loaded with growth factors such as bFGF, VEGF, PDGF on implantation into endodontically treated root canals after 3 weeks formed re-cellularized and revascularized pulp tissue with neodentin. Clinically, collagen sponge scaffold is effective in tooth production with maximum success. To manage periodontal and peri-implant defects, use of grafting material with collagen membranes is beneficial.

Fibrin has been used in bone/pulp tissue engineering because of biocompatibility, controlled biodegradability, and cells and biomolecules delivering capacity. Injectable 3D-shaped mold fibrin hydrogel is found to be suitable for pulp regeneration since it promotes functional well-organized structures. In vivo transplantation of polyethylene gylated fibrin hydrogel combined with DPSCs or PDLSCs revealed vascularized connective tissue resembling pulp tissue. Platelet-rich fibrin (PRF) with growth factors suggests complete regeneration of tooth with indiscriminate shape with crown, root, enamel, dentin, cementum, odontoblasts, pulp, blood vessels, and periodontal ligament. The PRF and HA graft material and PRF membrane are effective in management of large periapical lesion, whereas HA and PRF are effective in treatment of three-wall intrabony defects, and bone graft composed of PRF and β-tricalcium phosphate can be used to treat periapical cyst and for bone augmentation in periapical defects. The combination of PRF and mineral trioxide is biocompatible and favorable in managing pulpal floor perforation and in apexification. Investigations suggest that PRF on human dental pulpal cells enhances dental pulp cell proliferation, osteoprotegerin expression, and alkaline phosphatase activity. These findings suggest that PRF has a role in reparative dentin formation and in managing pulpitis.

Silk is a natural fibrous protein used in periodontal and maxillofacial therapies because of its biocompatibility, non-toxicity, physical characteristics, cell attachment, and cell proliferation. It also creates three-dimensional soft tissue augmentation. Hexafluoroisopropanol-based silk degrades slowly and supports soft dental pulp formation better than aqueous-based silk. Tooth bud cells when seeded on these scaffolds help in osteodentin formation. Micron-sized silk fibers when added produce high strength and serve as load-bearing bone grafts. They help in both human bone marrow stem cell differentiation and formation of bone-like tissue. Electrospun silkworm silk scaffolds allow gingival fibroblast attachment and proliferation of cells on electrospun fibers, thus promoting gingival tissue regeneration.

Polysaccharides

The commonly used natural polysaccharides for tooth regeneration are alginate, hyaluronic acid, agar, dextran, chitosan, and cellulose. Alginate is used as an alternative to proteins, polymers, and glass ceramics for bone/tooth regeneration. It is used in different forms like porous scaffold, injectable hydrogels, and nanofibrous scaffolds. RGD-modified alginate hydrogel helps in adhesion, proliferation, and differentiation of the cells. Alginate hydrogels promote dentin–pulp and periodontal regeneration by delivering TGF–β. Differentiation of odontoblast-like cells along with secretion of tubular dentin matrix is upregulated by TGF-β and acid-treated alginate hydrogels. Fibroblast cells from human periodontal ligament proliferate tremendously on alginate or bioglass scaffolds because nano-active bioglass ceramic has enhanced alkaline phosphatase activity. Hyaluronic acid is biocompatible and has low immunogenic activity so it offers as a preferable scaffold for dental pulp regeneration and blood vessel proliferation. The main disadvantage associated with this hyaluronic acid is in vivo rapid degradation and poor mechanical strength. Hyaluronic acid induces differentiation of mesenchymal cells and reparative dentin formation. Hyaluronic acid sponges and collagen sponges induce dental pulp proliferation with lesser inflammatory response [74]. Agar is a linear polysaccharide agarose which helps dental mesenchymal cells to differentiate into functional odontoblast-like cell, thus promoting tooth tissue formation [1]. Dextran is a complex polysaccharide, which is synthesized by sucrose, and helps in bone formation through the differentiation of bone forming cells. Hydrogel scaffold containing microspheres of dextran/gelatin loaded with BMP enhance periodontal tissue regeneration during periodontal therapy. Chitosan is usually processed into nanofibers, scaffolds, membranes, gels, beads, and sponges. It is a natural biopolymer which is derived from chitin used in dental applications from restorative dentistry to tissue engineered scaffolds for the alveolar bone to periodontal complex healing [75]. It shows high expression for BMP-2 osteoblast gene and increased alkaline phosphatase activity. Chitosan gel combined with demineralized bone matrix and the collagenous membrane has been used for periodontal regeneration, whereas HA chitosan scaffolds find application in bone tissue regeneration. The collagen tetracycline membrane also has beneficial effect on regeneration of bone and cementum in intrabony defects in animal models. Chitosan-based trilayer scaffold and solidified chitosan hydrogel have promising effect in periodontal regeneration by improving the functional ligament length [76]. Chitosan membranes loaded with growth factors like bFGF and PDGF-BB can induce more tissue regeneration and improves guided tissue regeneration. The blending of chitosan with other materials such as HA or silicone and preparation of chitosan nanofibers enhance the mechanical properties and make them appropriate for clinical application in guided tissue regeneration. Cellulose is a polysaccharide found in green plants, with excellent biocompatibility and better mechanical properties. Cellulose scaffolds exhibit angiogenic effect; hence, microfibrous cellulose acetate scaffolds improve the formation of a capillary tube-like structure, thus enhancing pulp revascularization and regeneration.

## 10.2  Synthetic Biomaterials

### 10.2.1  Ceramic Scaffold/Bioactive Ceramics

Ceramic scaffolds are calcium/phosphate materials ($\beta$-TCP or HA), bioactive glasses, and glass ceramics. They have varied benefits like biocompatibility, lesser immunogenicity, better healing of large bone defects, bone regeneration, and osteoconductivity. $\beta$-TCP and HA are two different entities, where $\beta$-TCP is biodegradable ceramic, whereas HA is non-biodegradable ceramic. $\beta$-TCP and biphasic calcium phosphate scaffold have been used for pulp and dentin tissue regeneration. 3D calcium phosphorus porous granules are used in dental tissue engineering as it provides 3D substrate conditions for the growth and differentiation of human dental pulp stem cell (hDP-SCs). Mechanical strength and cellular proliferation can be increased by the addition of silicon dioxide and zinc oxide. Calcium phosphate dissolution products increase the osteoblastic activity of the material. Disadvantages include poor mechanical strength, brittle nature, tough to shape, slower degradation rate, and high density. Glass ceramics-based $SiO_2$–$Na_2O$–$CaO$–$P_2O_5$ helps in osteogenic differentiation of pulp stem cells and provides good crystallization conditions. Ceramic scaffolds can be modified to enhance cellular activity. The mechanical integrity and bioactivity are enhanced during dental tissue regeneration by using Mg-based glass ceramics. Attachment, proliferation, and differentiation of hDPSCs are excellent using niobium-doped fluorapatite glass ceramics. Bioactive glass is considered as a good alternative to HA as they are able to bind to bone and as well as soft tissues. It can replace osseous defects and helps in developing new attachment on tooth surfaces. Dental ceramic coated with bioactive glass accelerates HA formation. Nanobioactive glass ceramic composite scaffold promotes maximum mineral deposition and is considered as a potential candidate for alveolar bone regeneration. They can be used as surface-reactive osteoconductive and osteoinductive biomaterials that promote osteogenesis and pro-angiogenesis and hence used in soft tissue engineering because of its high mechanical strength [77].

## 10.2.2   Synthetic Polymers

Polyesters

PGA, PLGA, PLLA, and polylactic acid (PLA) are commonly used synthetic polyester polymers for dental pulp regeneration. Polycaprolactone (PCL) is less commonly used because of limited bioactivity and extended degradation rate which can alter the molecular weight, crystallinity, and can modify the structure. PGA was first attempted in vitro for pulpal regeneration and revealed cellularity similar to the normal pulp tissue. PGA has better cell proliferation, conduction, increased regeneration of blood vessels, and fibroblast differentiation. PGA and PCL composite scaffolds seeded with genetically modified human cells lead to enhanced periodontal regeneration. PGA and PLA scaffolds were used for seeding SHED, DPSCs, and pulp fibroblasts to result into tissues similar to dentin and pulp by differentiation of stem cells into odontoblast-like cells and endothelial cells. PLLA can also show similar cellularity and pulp tissue formation. Combination of simvastatin and nanofibrous PLLA scaffold promotes the odontogenic potential of dental pulp cells in an inflammatory environment. A PGA/PLLA scaffold has shown to regenerate tooth crown composed of enamel, dentin, and pulp using tooth bud-isolated cells on transplanted rat omentum. The periodontal ligament tissues were also regenerated when similar constructs were translated into rat jaws. Another promising polymer named synthetic open-cell PLA was effective in dental pulp regeneration. Scaffold with SHED was able to adhere to the dentin of the root canal after transplantation into clean and shaped root canals of extracted human teeth.

Polyanhydrides

Polyanhydrides are surface-eroding polymers which contain two carbonyl groups linked by ether group. Polysebacic acid, poly 1, 3-bisp-carboxyphenoxypropane, and poly 1, 6-bis (p-carboxyphenoxy) hexane solely/blending with polymers (e.g., PCL) were explored for bone regeneration and repair. In hard tissue engineering, their use is restricted due to its unsatisfying mechanical properties particularly in load-bearing areas. Hence, to enhance the mechanical properties, the polyanhydrides are combined with polyimides as an alternative approach. In dentistry, polyanhydrides linked with carboxylic acid, amine, thiol, alcohol, or phenol group have gained attention due to its use in oral inflammatory pathologies [78, 79]. Similarly, Hasturk et al. [80] studied the utilization of light/chemically set polymethylmethacrylate, polyhydroxyethylmethacrylate, and calcium hydroxide graft material along with polyanhydride around dental implants and extraction sockets. The microscopic evaluations supported the implant stability with good bone-to-implant contact and well-organized implant–bone interface, effective in crestal augmentation throughout immediate implant placement. A modified perishable polyanhydride into disks, coatings, microspheres, and tubes was used for treatment for periodontal diseases, orthopedic injuries, nerve regeneration, and biofilm formation.

## 10.3  Composite Scaffold

Composite approach of combining natural or synthetic biomaterial (biopolymers and/or bioceramic) is a method to reduce the disadvantage and combine the advantages of each individual biomaterial. Composite materials often show excellent equilibrium in strength and toughness and improved characteristics compared to their individual components [81]. Hence, ceramic/polymer composite scaffolds have emerged. These composite scaffolds enhance the mechanical properties and osteoconductivity. Polymer–bioglass composite shows improvement in cell adhesion and growth of viable cells. Calcium phosphorus composite is more effective in tooth tissue regeneration, and PGLA/TCP scaffold is preferable for dentin–pulp regeneration. Bone reconstruction and regeneration can be done with zirconia hydroxyapatite composite scaffold as it has cellular/tissue compatibility and excellent mechanical properties.

# 11  Conclusions

The era of biomaterials and biomaterial engineering is focused toward improving the treatment outcomes of the patient. However, immense research is going on to regenerate dental tissues aiming for clinically appropriate structures and function. The application of stem cells and biomaterial sciences for tooth regeneration needs to be understood deeply. Since the current or emerging paradigms in tooth regeneration are showing limited and variable outcome with no true biological tissue formation, the translation of this tooth tissue engineering into clinical practices is still a question.

# References

1. Sharma S, Srivastava D, Grover S, Sharma V (2014) Biomaterials in tooth tissue engineering: a review. J Clin Diagn Res 8(1):309–315
2. Young CS, Terada S, Vacanti JP, Honda M, Bartlett JD, Yelick PC (2002) Tissue engineering of complex tooth structures on biodegradable polymer scaffolds. J Dent Res 81(10):695–700
3. Duailibi MT, Duailibi SE, Young CS, Bartlett JD, Vacanti JP, Yelick PC (2004) Bioengineered teeth from cultured rat tooth bud cells. J Dent Res 83(7):523–528
4. Honda MJ, Sumita Y, Kagami H, Ueda M (2005) Histological and immune histochemical studies of tissue engineered odontogenesis. Arch Histol Cytolo 68(2):89–101
5. Duailibi SE, Duailibi MT, Zhang W, Asrican R, Vacanti JP, Yelick PC (2008) Bioengineered dental tissues grown in the rat jaw. J Dent Res 87(8):745–750
6. Ahsan T (2007) Tissue engineering and regenerative medicine: advancing toward clinical therapies. Trans Approaches Tissue Eng Regener Med 3–16
7. Kim K, Lee CH, Kim BK, Mao JJ (2010) Anatomically shaped tooth and periodontal regeneration by cell homing. J Dent Res 89(8):842–847
8. Yuan Z, Nie H, Wang S, Lee CH, Li A, Fu SY, Zhou H, Chen L, Mao JJ (2011) Biomaterial selection for tooth regeneration. Tissue Eng Part B 17(5):373–388

9. Kim SG, Zheng Y, Zhou J, Chen M, Embree MC, Song K, Jiang N, Mao JJ (2013) Dentin and dental pulp regeneration by the patient's endogenous cells. Endod Top 28(1):106–117
10. Huang GJ, Garcia-Godoy F (2014) Missing concepts in de novo pulp regeneration. J Dent Res 93(8):717–724
11. Xiao L, Nasu M (2014) From regenerative dentistry to regenerative medicine: progress, challenges, and potential applications of oral stem cells. Stem Cells Cloning Adv Appl 7:89–99
12. Ohazama A, Modino SAC, Miletich I, Sharpe PT (2004) Stem-cell-based tissue engineering of murine teeth. J Dent Res 83(7):518–522
13. Arakaki M, Ishikawa M, Nakamura T, Iwamoto T, Yamada A, Fukumoto E, Saito M, Otsu K, Harada H, Yamada Y, Fukumoto S (2012) Role of epithelial-stem cell interactions during dental cell differentiation. J Biol Chem 287(13):10590–10601
14. Lee JH, Lee DS, Choung HW, Shon WJ, Seo BM, Lee EH, Cho JY, Park JC (2011) Odontogenic differentiation of human dental pulp stem cells induced by preameloblast-derived factors. Biomaterials 32(36):9696–9706
15. Nakao K, Morita R, Saji Y, Ishida K, Tomita Y, Ogawa M, Saitoh M, Tomooka Y, Tsuji T (2007) The development of a bioengineered organ germ method. Nat Methods 4(3):227
16. Ikeda E, Tsuji T (2008) Growing bioengineered teeth from single cells: potential for dental regenerative medicine. Expert Opin Biol Ther 8(6):735–744
17. Ikeda E, Morita R, Nakao K, Ishida K, Nakamura T, Takano-Yamamoto T, Ogawa M, Mizuno M, Kasugai S, Tsuji T (2009) Fully functional bioengineered tooth replacement as an organ replacement therapy. Proc Natl Acad Sci 106(32):13475–13480
18. Oshima M, Mizuno M, Imamura A, Ogawa M, Yasukawa M, Yamazaki H, Morita R, Ikeda E, Nakao K, Takano-Yamamoto T, Kasugai S (2011) Functional tooth regeneration using a bioengineered tooth unit as a mature organ replacement regenerative therapy. PloS One 6(7):p.e21531
19. Hartgerink JD, Beniash E, Stupp SI (2001) Self-assembly and mineralization of peptide-amphiphile nanofibers. Science 294(5547):1684–1688
20. Xu R, Zhou Y, Zhang B, Shen J, Gao B, Xu X, Ye L, Zheng L, Zhou X (2015) Enamel regeneration in making a bioengineered tooth. Curr Stem Cell Res Ther 10(5):434–442
21. Yu JH, Shi JN, Deng ZH, Zhuang H, Nie X, Wang RN, Jin Y (2006) Cell pellets from dental papillae can reexhibit dental morphogenesis and dentinogenesis. Biochem Biophys Res Commun 346(1):116–124
22. Iohara K, Nakashima M, Ito M, Ishikawa M, Nakasima A, Akamine A (2004) Dentin regeneration by dental pulp stem cell therapy with recombinant human bone morphogenetic protein. J Dent Res 83(8):590–595
23. Fukumoto S, Miner JH, Ida H, Fukumoto E, Yuasa K, Miyazaki H, Hoffman MP, Yamada Y (2006) Laminin α5 is required for dental epithelium growth and polarity and the development of tooth bud and shape. J Biol Chem 281(8):5008–5016
24. Liu L, Liu YF, Zhang J, Duan YZ, Jin Y (2016) Ameloblasts serum free conditioned medium: bone morphogenic protein 4 induced odontogenic differentiation of mouse induced pluripotent stem cells. J Tissue Eng Regener Med 10(6):466–474
25. Zheng LW, Linthicum L, DenBesten PK, Zhang Y (2013) The similarity between human embryonic stem cell-derived epithelial cells and ameloblast-lineage cells. Int J Oral Sci 5(1):1–6
26. He P, Zhang Y, Kim SO, Radlanski RJ, Butcher K, Schneider RA, Den-Besten PK (2010) Ameloblast differentiation in the human developing tooth: effects of extracellular matrices. Matrix Biol 29(5):411–419
27. Kollar EJ, Fisher C (1980) Tooth induction in chick epithelium: expression of quiescent genes for enamel synthesis. Science 207(4434):993–995
28. Bing Hu, Nadiri Amal, Bopp Sabine Kuchler (2006) Tissue engineering of tooth crown, root, and periodontium. Tissue Eng Part A 12(8):2069–2075
29. Storrie H, Guler MO, Abu-Amara SN, Volberg T, Rao M, Geiger B, Stupp SI (2007) Supramolecular crafting of cell adhesion. Biomaterials 28(31):4608–4618
30. Stupp SI, Hartgerink JD, Beniash E, Northwestern University (2009) Self-assembly and mineralization of peptide-amphiphile nanofibers. US Patent 7,491,690

31. Huang Z, Sargeant TD, Hulvat JF, Mata A, Bringas-Jr P, Koh CY, Stupp SI, Snead ML (2008) Bioactive nanofibers instruct cells to proliferate and differentiate during enamel regeneration. J Bone Miner Res 23(12):1995–2006
32. Borovjagin AV, Dong J, Passineau MJ, Ren C, Lamani E, Mamaeva OA, Wu H, Keyser E, Murakami M, Chen S, MacDougall M (2011) Adenovirus gene transfer to amelogenesis imperfecta ameloblast-like cells. PloS One 6(10):p.e24281
33. Sharp T, Wang J, Li X, Cao H, Gao S, Moreno M, Amendt BA (2014) A pituitary homeobox 2 (Pitx2): microRNA-200a-3p: beta-catenin pathway converts mesenchyme cells to amelogenin-expressing dental epithelial cells. J Biol Chem 289(39):27327–27341
34. Akhter M, Kobayashi I, Kiyoshima T, Matsuo K, Yamaza H, Wada H, Honda JY, Ming X, Sakai H (2005) Possible functional involvement of thymosin beta 4 in developing tooth germ of mouse lower first molar. Histochem Cell Biol 124(3–4):207–213
35. Shiotsuka M, Wada H, Kiyoshima T, Nagata K, Fujiwara H, Kihara M, Hasegawa K, Someya H, Takahashi I, Sakai H (2014) The expression and function of thymosin beta 10 in tooth germ development. Int J Dev Biol 57(11–12):873–883
36. Wang B, Li L, Du S, Liu C, Lin X, Chen Y, Zhang Y (2010) Induction of human keratinocytes into enamel-secreting ameloblasts. Dev Biol 344(2):795–799
37. Inoue T (1986) Induction of chondrogenesis in muscle, skin, bone marrow, and periodontal ligament by demineralaized dentin and bone matrix in vivo and in vitro. J Dent Res 65:15–21
38. Bessho K, Tanaka N, Matsumoto J, Tagawa T, Murata M (1991) Human dentin-matrix-derived bone morphogenetic protein. J Dent Res 70(3):171–175
39. Rutherford RB, Wahle J, Tucker M, Rueger D, Charette M (1993) Induction of reparative dentine formation in monkeys by recombinant human osteogenic protein-1. Arch Oral Biol 38(7):571–576
40. Shinmura Y, Tsuchiya S, Hata KI, Honda MJ (2008) Quiescent epithelial cell rests of Malassez can differentiate into ameloblast like cells. J Cell Physiol 217(3):728–738
41. Murray PE, Garcia-Godoy F, Hargreaves KM (2007) Regenerative endodontics: a review of current status and a call for action. J Endod 33(4):377–390
42. Fouad AF (2011) The microbial challenge to pulp regeneration. Adv Dent Res 23(3):285–289
43. Sun HH, Jin T, Yu Q, Chen FM (2011) Biological approaches toward dental pulp regeneration by tissue engineering. J Tissue Eng Regener Med 5(4):e1–e16
44. Morse DR, O'Larnic J, Yesilsoy C (1990) Apexification: review of the literature. Quintessence Int 21(7):58–598
45. Suzuki T, Lee CH, Chen M, Zhao W, Fu SY, Qi JJ, Chotkowski G, Eisig SB, Wong A, Mao JJ (2011) Induced migration of dental pulp stem cells for in vivo pulp regeneration. J Dent Res 90(8):1013–1018
46. Pan S, Dangaria S, Gopinathan G, Yan X, Lu X, Kolokythas A, Niu Y, Luan X (2013) SCF promotes dental pulp progenitor migration, neovascularization, and collagen remodeling-potential applications as a homing factor in dental pulp regeneration. Stem Cell Rev Rep 9(5):655–667
47. Takeuchi N, Hayashi Y, Murakami M, Alvarez FJ, Horibe H, Iohara K, Nakata K, Nakamura H, Nakashima M (2015) Similar in vitro effects and pulp regeneration in ectopic tooth transplantation by basic fibroblast growth factor and granulocyte colony stimulating factor. Oral Dis 21(1):113–122
48. Yang JW, Zhang YF, Wan CY, Sun ZY, Nie S, Jian SJ, Zhang L, Song GT, Chen Z (2015) Autophagy in SDF-1α-mediated DPSC migration and pulp regeneration. Biomaterials 44:11–23
49. Zhang LX, Shen LL, Ge SH, Wang LM, Yu XJ, Xu QC, Yang PS, Yang CZ (2015) Systemic BMSC homing in the regeneration of pulp-like tissue and the enhancing effect of stromal cell-derived factor-1 on BMSC homing. Int J Clin Exp Pathol 8(9):10261
50. Li L, Wang Z (2016) PDGF-BB, NGF and BDNF enhance pulp-like tissue regeneration via cell homing. RSC Adv 6(111):109519–109527
51. Wang HL, Greenwell H, Fiorellini J, Giannobile W, Offenbacher S, Salkin L, Townsend C, Sheridan P, Genco RJ (2005) Periodontal regeneration. J Periodontol 76(9):1601–1622

52. Gronthos S, Mankani M, Brahim J, Robey PG, Shi S (2000) Postnatal human dental pulp stem cells (DPSCs) in vitro and in vivo. Proc Natl Acad Sci 97(25):13625–13630
53. Shi S, Bartold PM, Miura M, Seo BM, Robey PG, Gronthos S (2005) The efficacy of mesenchymal stem cells to regenerate and repair dental structures. Orthod Craniofac Res 8(3):191–199
54. Gronthos S, Brahim J, Li W, Fisher LW, Cherman N, Boyde A, DenBesten P, Robey PG, Shi S (2002) Stem cell properties of human dental pulp stem cells. J Dent Res 81(8):531–535
55. Koyama N, Okubo Y, Nakao K, Bessho K (2009) Evaluation of pluripotency in human dental pulp cells. J Oral Maxillofac Sur 67(3):501–506
56. Bressan E, Ferroni L, Gardin C, Pinton P, Stellini E, Botticelli D, Sivolella S, Zavan B (2012) Donor age-related biological properties of human dental pulp stem cells change in nanostructured scaffolds. PLoS One 7(11):49146
57. d'Aquino R, Graziano A, Sampaolesi M, Laino G, Pirozzi G, De Rosa A, Papaccio G (2007) Human postnatal dental pulp cells co-differentiate into osteoblasts and endotheliocytes: a pivotal synergy leading to adult bone tissue formation. Cell Death Differ 14(6):1162
58. Yu J, He H, Tang C, Zhang G, Li Y, Wang R, Shi J, Jin Y (2010) Differentiation potential of STRO-1+ dental pulp stem cells changes during cell passaging. BMC Cell Biol 11(1):32
59. Kaushik SN, Kim B, Walma AMC, Choi SC, Wu H, Mao JJ, Jun HW, Cheon K (2016) Biomimetic microenvironments for regenerative endodontics. Biomater Res 20(1):14
60. Wongwatanasanti N, Jantarat J, Sritanaudomchai H, Hargreaves KM (2018) Effect of bioceramic materials on proliferation and odontoblast differentiation of human stem cells from the apical papilla. J Endod 44(8):1270–1275
61. Hong S, Chen W, Jiang B (2018) A comparative evaluation of concentrated growth factor and platelet-rich fibrin on the proliferation, migration, and differentiation of human stem cells of the apical papilla. J Endod 44(6):977–983
62. Seo BM, Miura M, Gronthos S, Bartold PM, Batouli S, Brahim J, Young M, Robey PG, Wang CY, Shi S (2004) Investigation of multipotent postnatal stem cells from human periodontal ligament. Lancet 364(9429):149–155
63. Yoo JH, Lee SM, Bae MK, Lee DJ, Ko CC, Kim YI, Kim HJ (2018) Effect of orthodontic forces on the osteogenic differentiation of human periodontal ligament stem cells. J Oral Sci 17:0310
64. Wang Y, Zhou Y, Jin L, Pang X, Lu Y, Wang Z, Yu Y, Yu J (2018) Mineral trioxide aggregate enhances the osteogenic capacity of periodontal ligament stem cells via NFκB and MAPK signaling pathways. J Cell Physiol 233(3):2386–2397
65. Yan XZ, Beucken JJ, Yuan C, Jansen JA, Yang F (2018) Spheroid formation and stemness preservation of human periodontal ligament cells on chitosan films. Oral Dis 24(6):1083–1092
66. Handa K, Saito M, Yamauchi M, Kiyono T, Sato S, Teranaka T, Narayanan AS (2002) Cementum matrix formation in vivo by cultured dental follicle cells. Bone 31(5):606–611
67. Guo L, Li J, Qiao X, Yu M, Tang W, Wang H, Guo W, Tian W (2013) Comparison of odontogenic differentiation of human dental follicle cells and human dental papilla cells. PLoS One 8(4):62332
68. Batouli S, Miura M, Brahim J (2003) Comparison of stem-cell- mediated osteogenesis and dentinogenesis. J Dent Res 82:976–981
69. Cordeiro MM, Dong Z, Kaneko T, Zhang Z, Miyazawa M, Shi S, Smith AJ, Nor JE (2008) Dental pulp tissue engineering with stem cells from exfoliated deciduous teeth. J Endod 34(8):962–969
70. Nakamura S, Yamada Y, Katagiri W, Sugito T, Ito K, Ueda M (2009) Stem cell proliferation pathways comparison between human exfoliated deciduous teeth and dental pulp stem cells by gene expression profile from promising dental pulp. J Endod 35(11):1536–1542
71. Albrektsson T, Johansson C (2001) Osteoinduction, osteoconduction and osseointegration. Eur Spine J 10(2):S96–S101
72. Bhui AS, Singh G, Sidhu SS, Bains PS (2018) Experimental investigation of optimal ED machining parameters for Ti-6Al-4V biomaterial. FU Ser Mech Eng 16(3):337–345
73. Granito RN, Custodio MR, Renno ACM (2017) Natural marine sponges for bone tissue engineering: the state of art and future perspectives. J Biomed Mater Res Part B Appl Biomater 105(6):1717–1727

74. Ganesh N, Hanna C, Nair SV, Nair LS (2013) Enzymatically cross-linked alginic-hyaluronic acid composite hydrogels as cell delivery vehicles. Int J Biol Macromol 55:289–294
75. Husain S, Al-Samadani KH, Najeeb S, Zafar MS, Khurshid Z, Zohaib S, Qasim SB (2017) Chitosan biomaterials for current and potential dental applications. Materials 10(6):602
76. Varoni EM, Vijayakumar S, Canciani E, Cochis A, De-Nardo L, Lodi G, Rimondini L, Cerruti M (2018) Chitosan-based trilayer scaffold for multitissue periodontal regeneration. J Dent Res 97(3):303–311
77. Srinivasan S, Jayasree R, Chennazhi KP, Nair SV, Jayakumar R (2012) Biocompatible alginate/nano bioactive glass ceramic composite scaffolds for periodontal tissue regeneration. Carbohydr Polym 87(1):274–283
78. Uhrich KE, Rutgers State University of New Jersey (2010) Polyanhydride linkers for production of drug polymers and drug polymer compositions produced thereby. US Patent 7,666,398
79. Conte R, Di-Salle A, Riccitiello F, Petillo O, Peluso G, Calarco A (2018) Biodegradable polymers in dental tissue engineering and regeneration. Mater Sci 5(6):1073–1101
80. Hasturk H, Kantarci A, Ghattas M, Dangaria SJ, Abdallah R, Morgan EF, Diekwisch TG, Ashman A, Van-Dyke T (2014) The use of light/chemically hardened polymethylmethacrylate, polyhydroxylethylmethacrylate, and calcium hydroxide graft material in combination with polyanhydride around implants and extraction sockets in minipigs: part II: histologic and micro CT evaluations. J Periodontol 85(9):1230–1239
81. Bains PS, Sidhu SS, Payal HS (2016) Fabrication and machining of metal matrix composites: a review. Mater Manuf Process 31(5):553–573

# Chapter 8
# Surface Characteristics and In Vitro Corrosion Behavior of HAp-coated 316L Stainless Steel for Biomedical Applications

**Gurpreet Singh, Amandeep Singh Bhui, Sarabjeet Singh Sidhu, Preetkanwal Singh Bains and Yubraj Lamichhane**

## 1 Introduction

The study of biomaterials is going on since many decades. The need for safe and efficient biomaterials is rising tremendously because of increase in number of accidents causing fractured human bones, joint replacement, etc. The basic requirement of biomedical implants is the existences of proper physiological environment inside the body, strength, wear resistance, biological fixation, cell growth and corrosion resistance [1]. The surface interface and its properties determine the acceptance or rejection of the biomedical implants on the body [2–4].

The quality of machined surface highly affects the bioactive properties like corrosion resistance, wear resistance and cell proliferation of the metallic bio-implants [5, 6]. Success of biomaterials is based on the characteristics such as topography, chemical composition, mechanical strength, corrosion resistance and surface roughness.

G. Singh (✉) · S. S. Sidhu · P. S. Bains
Department of Mechanical Engineering, Beant College of Engineering and Technology, Gurdaspur 143521, India
e-mail: singh.gurpreet191@gmail.com

S. S. Sidhu
e-mail: sarabjeetsidhu@yahoo.com

P. S. Bains
e-mail: preetbains84@gmail.com

A. S. Bhui · Y. Lamichhane
Department of Mechanical Engineering, Amritsar College of Engineering and Technology, Amritsar 143109, India
e-mail: meet_amandeep@yahoo.com

Y. Lamichhane
e-mail: lamichhaneyubraj999@gmail.com

© Springer Nature Singapore Pte Ltd. 2019
P. S. Bains et al. (eds.), *Biomaterials in Orthopaedics and Bone Regeneration*,
Materials Horizons: From Nature to Nanomaterials,
https://doi.org/10.1007/978-981-13-9977-0_8

117

Hence, many surface modification and surface coating techniques are being used nowadays to enhance the required properties of biomedical implants. Modifications of biomaterials also overcome the risk of rejection due to loosening of joint, toxicity, infection within the individual providing proper cell adhesion, cell proliferations and antibacterial properties [7–10].

Among all of the biomaterials, i.e., polymeric, ceramic, natural, metallic and composites, metallic biomaterials are widely acceptable due to their mechanical properties and compatibility with the human bone. Titanium alloys, 316L stainless steel, chromium and cobalt are the metallic biomaterials; amid all, 316L stainless steel because of their easy availability, low cost, better adhesion and cell growth preferred much more [11–15]. But the presence of sodium, chlorine, water, saliva and amino acids within the human body disturbs the equilibrium state and consumes the metallic implant by various anodic and/or cathodic reactions [16–18]. As a result, metallic biomaterial implant corrodes after certain years of implantation. In order to overcome such issue, coating of bioactive layer on surface is best solution which not only evades early corrosion but also avoids release of harmful ions [19].

Electric discharge coating (EDC) is an application of well-known electrical discharge machine (EDM) by reversing the polarity during the machining process. Reverse polarity alters the flow of electrons compared to conventional process and deposits the material mixed in dielectric to modify the surface characteristics of the workpiece [20–23]. The intermetallic compounds and carbides thus formed enhance the properties of substrate biomaterial offering better adhesion and proliferation of human osteoblast-like MG-63 cells.

## 2 Literature Review

Among numerous studies on metallic biomaterials, Mahajan and Sidhu [24] reviewed the need for surface modification of biomaterials for the enhancement of their functionality. The surface modification of metallic biomaterials, i.e., stainless steel, titanium using EDM, helps to obtain the biocompatible surfaces, increases bond ability, etc. EDM is the method that has the ability to replicate the architecture and characteristics of the natural bones. It had great applications in biomedical fields for the enrichment of required surface characteristics. Other researchers like Prakash et al. [25] also expressed EDM as a potential and new innovative way for surface modification of various metallic implants for orthopedic applications. According to the in vitro bioactivity analysis, powder-mixed dielectric machined surface offers better cell growth than substrate and surface machined without the addition of powder in dielectric [26].

The surface of 316L stainless steel was treated by Mazur et al. [27] utilizing sol–gel technique to deposit ceramic layer of $SiO_2–Y_2O_3$. Newly formed ceramic layer helped to enhance the bioactivity and corrosion resistance of 316L which was confirmed by EDS and Raman spectra analysis. Compounds like calcium (Ca) and phosphorous (P) were observed confirming the formation of apatite ceramic layer on stainless steel 316L with improved biocompatibility. Electron beam physical vapor deposition (EBPVD) was employed by Kaliaraj et al. [28] to modify 316L stainless steel surface with monoclinic ($m$-$ZrO_2$) and tetragonal ($t$-$ZrO_2$) phase of zirconium dioxide. They found improved cell viability during the cell culture analysis of coated samples. Superior corrosion resistance was shown by coated samples of 316L stainless steel in artificial blood plasma (ABP) as electrolyte using electrochemical test.

Abbas et al. [29] investigated the present research scenario of electrical discharge machining, suggesting PMEDM as dominating non-conventional machining process for the improvement of surface characteristics with the addition of powder in dielectric medium. Here, high MMR, better surface finish and low TWR can be obtained. The surface characteristics of aluminum powder-mixed ED machined Ti-6Al-4V was examined by Abdul-Rani et al. [30] for the formation of intermetallic compounds and carbon enriched layer on the substrate surface which helps in osseointegration and bone proliferations. Furthermore, modified surface exhibited enhanced corrosion resistance of titanium alloy.

Singh et al. [31] investigated the effect of input process parameters on MRR by machining AISI 316L stainless steel using EDM process. Current was found as the most significant factor influencing MRR. Highest MRR of 6.25 mg/min was obtained at the combination of current 28 A, $T_{on}$ 90 µs, $T_{off}$ 60 µs and voltage 80 V. The performance of traditional EDM process with the addition of Al powder using reverse polarity was studied by Sharma et al. [32]. Their investigation showed that various characteristics of powder like concentration, size, etc., had its dominating effect on the performance of electric discharge machine.

Karamian et al. [33] coated HAp/zircon on stainless steel 316L and concluded that powder coating helps to improve adhesion properties of metallic biomaterial. The investigation of HAp/$TiO_2$ coating on the titanium substrate was performed by Ramires et al. [34] for improving the bioactive properties such as osseointegration and cell proliferations. Manam et al. [35] intensely reviewed the biocompatibility of metallic biomaterials, i.e., titanium alloys, stainless steel and chromium–cobalt alloys. They studied that corrosion resistance of biomaterial was an important factor. Moreover, corrosive nature causes the failure of implantation and leads to surgery. However, surface modification of biomaterials with bioactive layer offered better corrosion resistance and evaded the harmful ions to release. They concluded that metallic biomaterials could not be replaced with polymers and ceramics due to the required mechanical properties for a material to be biomaterial.

Bains et al. [36] and Long et al. [37] optimized the MRR by employing PMEDM on using copper and graphite tool with titanium and SiC-mixed dielectric. They concluded that powder concentration increases metal erosion as compared to no powder in dielectric with discharge current as one of the most significant factors followed by other variables.

In this research, EDC was employed on 316L stainless steel with hydroxyapatite nanoparticles-mixed dielectric to examine improved surface characteristics in terms of surface roughness and formation of bioactive compounds on 316L surface. Additionally, in vitro corrosion analysis was performed on machined province to scrutinize the improved corrosion resistance of HAp-coated 316L stainless steel surface.

# 3 Methodology

## 3.1 Materials

Medical grade 316L stainless steel was procured from Metline Industries, Mumbai, for the experimentation in the form of rectangular plate (size: 50 mm × 100 mm × 5 mm). Electrolytic copper tool because of its conductivity with dia. 900 μm was chosen as electrode for the ED machining of 316L SS. The compositional analysis of 316L stainless steel is listed in Table 1.

Hydroxyapatite nanopowder with average particle size 20–45 nm and purity of 99.5% mixed in dielectric medium for the surface modification of 316L stainless steel.

## 3.2 Design of Experiment

Taguchi's orthogonal arrays are generally executed to reduce the number of experimental trials according to selected machining parameters. In the present work, $L_{18}$ $(2^1 \times 3^4)$ was used for conducting the experiments selecting five input parameters, i.e., dielectric medium, discharge current, pulse-on-time, pulse-off-time and voltage. Minitab-17 was utilized to generate design of experiments with three levels of input parameter as shown in Table 2.

**Table 1** Chemical composition of 316L stainless steel

| Element | C | Mn | Si | Cr | Mo | Ni | S | N | P | Fe |
|---------|------|------|------|-------|------|-------|------|------|-------|---------|
| % | 0.03 | 2.00 | 0.75 | 16.00 | 2.00 | 10.00 | 0.03 | 0.10 | 0.045 | Balance |

**Table 2** Parameters and their levels

| Input parameters | Units | Level 1 | Level 2 | Level 3 |
|------------------|-----------|---------|---------------|---------|
| Dielectric medium | – | EDM oil | EDM oil + HAp | – |
| Discharge current | Ampere | 20 | 24 | 28 |
| Pulse-on-time | μ-seconds | 60 | 90 | 120 |
| Pulse-off-time | μ-seconds | 60 | 90 | 120 |
| Voltage | Volts | 40 | 60 | 80 |

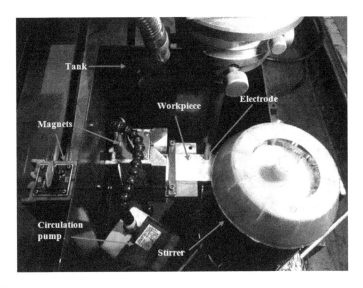

**Fig. 1** Experimental setup

## 3.3  Experimentation

Die sinker-type electrical discharge machine with reverse polarity {workpiece as cathode (−) and electrode as anode (+)} was employed for conducting the experimentation. Out of 18 experimental runs, nine were performed in pure dielectric, i.e., EDM oil, whereas for powder-mixed experimentation, indigenously developed setup (shown in Fig. 1) consisting of stirrer and pump for proper and homogeneous mixing of HAp particles was used. Machining time for each run was fixed for 40 min with powder concentration of 15 g/l. After the ED machining, Mitutoyo SJ-400 surface roughness tester was used to measure roughness ($R_a$) of machined surface. Each sample measured diametrically from three locations and was averaged for further analysis.

In this experimental investigation, dielectric medium, discharge current ($I$), pulse-on-time ($T_{on}$), pulse-off-time ($T_{off}$) and Voltage ($V$) were selected as input parameters being the most common and influencing parameters during the machining process. The selected output response was surface roughness ($R_a$) in μm.

## 3.4  In Vitro Corrosion Analysis

Electrochemical corrosion test was performed to validate the enhanced corrosion resistance and subsequently low corrosion rate of modified 316L stainless steel surface with HAp particles. Ringer solution at 37 °C was used as simulated body fluid (SBF) to replicate the human body environment as electrolyte immersing 316L sample as working electrode (WE), whereas platinum wire as counter electrode (CE) and Ag/AgCl employed as reference electrode (RE). Prior to analysis, specimen was bathed in acetone and insulated with tape leaving an exposed machined area of 0.32 cm$^2$. The insulation restricts the contribution of base metal during the electrochemical testing.

# 4 Results and Discussion

The experimental design matrix and surface roughness of machined specimen are shown in Table 3. The surface response table is represented according to signal-to-noise ratio (S/N) methodology—a ratio of strength of signal to the magnitude of error. S/N ratios depend on the type of responses measured such as "Larger is better" type which is given by the equation below:

$$(S/N)_{Larger} = -10 \log \left\{ \frac{1}{R} \sum_{i=1}^{R} \left( \frac{1}{y_i^2} \right) \right\}$$

where $R$ = number of repetitions of responses; $y_i$ = value of response at $i$th trial.

## 4.1 Analysis of Variance for S/N Ratios of Surface Roughness

The surface response of machined workpiece was analyzed using analysis of variance (ANOVA) of S/N ratios as shown in Table 4 with the aid of Minitab-17 software. Momentous process parameters affecting the roughness of machined 316L stainless steel were recognized using p-value and subsequently their percentage contribution. Pulse-on-time was the most dominating parameter affecting surface roughness (contribution 33.40%) followed by discharge current (contribution 24.52%) and dielectric medium (contribution 16.81%). Consequently, the desired porous surface for proper adhesion, cell growth validating the surface modification of selected biomaterial examined in HAp powder-mixed dielectric at higher value of on-time (i.e., 90 μs) and peak current (i.e., 28 A) which concurrently endow with the desired surface topography and phase transformation of 316L stainless steel for requisite bioactivity. Figure 2 illustrates the S/N ratio plot for surface roughness.

Further, powder-mixed ED machined 316L stainless steel surface was examined for changed morphology, formation of intermetallic compounds and in vitro corrosion analysis to validate the surface modification with bioactive powder and enhanced bioactivity.

**Table 3** Experimental design and surface roughness response

| Exp. No. | Dielectric medium | $I$ (A) | $T_{on}$ (μs) | $T_{off}$ (μs) | Voltage (V) | Surface roughness (μm) | | | S/N ratio |
|---|---|---|---|---|---|---|---|---|---|
| | | | | | | R1 | R2 | R3 | |
| 1 | EDM oil | 20 | 60 | 60 | 40 | 0.46 | 0.15 | 0.72 | −12.3129 |
| 2 | EDM oil | 20 | 90 | 90 | 60 | 0.27 | 0.41 | 0.73 | −8.5618 |
| 3 | EDM oil | 20 | 120 | 120 | 80 | 0.75 | 0.87 | 0.86 | −1.7134 |
| 4 | EDM oil | 24 | 60 | 60 | 60 | 0.44 | 0.62 | 0.66 | −5.2558 |
| 5 | EDM oil | 24 | 90 | 90 | 80 | 0.43 | 0.24 | 0.19 | −12.2591 |
| 6 | EDM oil | 24 | 120 | 120 | 40 | 1.12 | 1.01 | 0.96 | 0.2037 |
| 7 | EDM oil | 28 | 60 | 90 | 40 | 0.59 | 0.76 | 0.82 | −3.0759 |
| 8 | EDM oil | 28 | 90 | 120 | 60 | 0.51 | 0.49 | 0.64 | −5.4204 |
| 9 | EDM oil | 28 | 120 | 60 | 80 | 1.09 | 1.18 | 1.22 | 1.2848 |
| 10 | EDM oil + HAp powder | 20 | 60 | 120 | 80 | 0.58 | 0.50 | 0.78 | −4.5820 |
| 11 | EDM oil + HAp powder | 20 | 90 | 60 | 40 | 0.26 | 1.08 | 0.93 | −7.4835 |
| 12 | EDM oil + HAp powder | 20 | 120 | 90 | 60 | 1.04 | 1.24 | 0.87 | 0.1531 |
| 13 | EDM oil + HAp powder | 24 | 60 | 90 | 80 | 0.69 | 0.94 | 0.54 | −3.4645 |
| 14 | EDM oil + HAp powder | 24 | 90 | 120 | 40 | 0.56 | 0.68 | 0.75 | −3.7592 |
| 15 | EDM oil + HAp powder | 24 | 120 | 60 | 60 | 0.89 | 1.13 | 1.19 | 0.3751 |
| 16 | EDM oil + HAp powder | 28 | 60 | 120 | 60 | 1.37 | 1.26 | 0.77 | 0.2238 |
| 17 | EDM oil + HAp powder | 28 | 90 | 60 | 80 | 1.76 | 1.29 | 1.52 | 3.4468 |
| 18 | EDM oil + HAp powder | 28 | 120 | 90 | 40 | 1.25 | 0.87 | 1.21 | 0.5483 |

*R* Repetitions

**Table 4** Analysis of variance for S/N ratios of surface roughness

| Source | DF | Seq SS | Adj SS | Adj MS | $p$-value | % contribution |
|---|---|---|---|---|---|---|
| Dielectric medium | 1 | 58.929 | 58.929 | 58.929 | $0.032^*$ | 16.81 |
| Discharge current (A), $I$ | 2 | 85.986 | 85.986 | 42.993 | $0.041^*$ | 24.52 |
| Pulse-on-time ($\mu$s), $T_{on}$ | 2 | 117.103 | 117.103 | 58.551 | $0.020^*$ | 33.40 |
| Pulse-off-time ($\mu$s), $T_{off}$ | 2 | 11.329 | 11.329 | 5.665 | 0.549 | 3.24 |
| Voltage (V) | 2 | 7.218 | 7.218 | 3.609 | 0.676 | 2.05 |
| Residual error | 8 | 70.083 | 70.083 | 8.760 | | 19.98 |
| Total | 17 | 350.648 | | | | 100 |

*Significant at 99% confidence level, Rank 1: pulse-on, Rank 2: current, Rank 3: dielectric

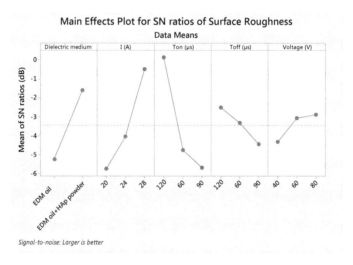

**Fig. 2** Main effects plot for S/N ratios of surface roughness

## 4.2 Surface Morphology of Machined Surface

Scanning electron microscopy (JSM-6610 LV Joel, Japan) was used to inspect the morphology of machined 316L specimen in hydroxyapatite nanopowder-mixed dielectric medium. Deposition of powder particles and presence of homogeneous porous structure surface encouraging cell growth, adhesion between bone and implant, and cell proliferation were revealed by SEM (Fig. 3) analysis.

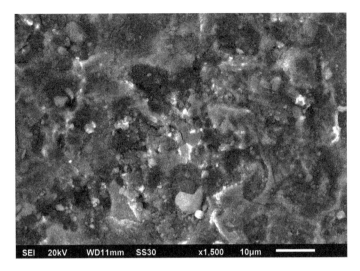

**Fig. 3** Surface representing porous structure at $I = 28$ A; $T_{on} = 90$ μs; $T_{off} = 60$ μs; voltage = 80 V

## 4.3  Phase Transformation of Machined Surface

It was described in the previous research [38, 39] that sparks produced during ED machining changed the chemical composition of the machined surface. X-ray diffraction (XRD) technique was used to investigate the formation of intermetallic and bioactive compounds on ED machined 316L stainless steel surface (trial 17). Figure 4 witnesses the XRD pattern confirming the formation and deposition of intermetallic compounds, carbides as well as silicides on the EDMed surface. Newly formed

**Fig. 4**  XRD pattern showing formation of new compounds

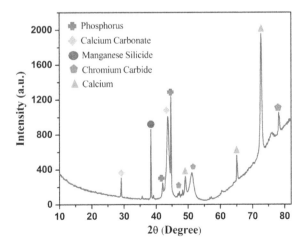

**Fig. 5** Tafel plot showing
polarization curve of
specimens (trial 17)

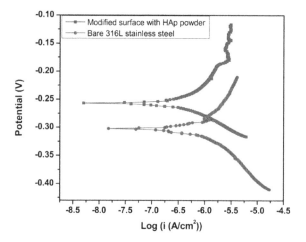

bioactive compounds contribute for enhanced bioactivity in terms of improved corrosion resistance, proper adhesion between bone and implant, and cell growth.

## 4.4   In Vitro Corrosion Analysis

After the examination of porous structure and compositional analysis using SEM and
XRD, respectively, electrochemical corrosion analysis was carried out to validate
the response of modified surface in terms of enhanced corrosion resistance. Ringer's
solution with pH value 7.2 at 37 °C was used as simulated body fluid (SBF) to
reproduce the human body environment for the current investigation.

It was evident from the Tafel plot (Fig. 5) that surface modified with HAp powder-
mixed dielectric depicts higher value of corrosion potential ($E_{corr} = -257.810$ mV)
compared to bare metal possess $-308.490$ mV. Consequently, modified surface of
316L stainless steel showed improved corrosion rate of 0.0972 mm/year compared
to substrate material with corrosion rate of 1.79 mm/year. The newly modified ED
machined 316L surface exhibited improved corrosion resistance that is necessitated
for better and proper bone–implant adhesion. It will avoid the release of metallic
ions due to the presence of enzymes and reacting environment within the individual
causing poor osseointegration and cytotoxicity.

## 5   Conclusions

Present work analyzed the surface modification of medical grade 316L stainless steel
using electric discharge coating (EDC) in HAp powder-mixed dielectric. Based on

the surface characteristics and in vitro corrosion analysis, following conclusions were drawn:

1. Surface roughness directly increased with the increase in pulse-on and current applied. Desired porous surface (1.523 μm) was observed at 28A; $T_{on}$ 90 μs; $T_{off}$ 60 μs; and 80 V in the presence of HAp in the dielectric medium.
2. SEM revealed porosity and deposition of powder particles on HAp powder-mixed dielectric machined surface.
3. XRD confirmed the phase transformation of surface (trial 17) with the formation of bioactive compounds (calcium, phosphorus, calcium carbonate) and several intermetallic compounds.
4. Surface machined at higher discharge current, i.e., 28 A sounds more crowded with intermetallic compounds and carbides compared to surface machined at lower value of discharge current.
5. In vitro corrosion analysis exhibited improved corrosion resistance and accordingly low corrosion rate of HAp powder modified 316L stainless steel surface (0.0972 mm/year) compared to bare metal (1.79 mm/year).
6. Enhanced surface characteristics and corrosion resistance validate the surface modification of 316L with HAp powder using EDC for biomedical applications.

# References

1. Bhui AS, Singh G, Sidhu SS, Bains PS (2018) Experimental investigation of optimal ED machining parameters for Ti-6Al-4V biomaterial. FU Ser Mech Eng 16(3):337–345
2. Mahajan A, Sidhu SS (2018) Enhancing biocompatibility of Co-Cr alloy implants via electrical discharge process. Mater Technol 33(8):524–531
3. Tasnim N, Kumar A, Joddar B (2017) Attenuation of the in vitro neurotoxicity of 316L SS by graphene oxide surface coating. Mater Sci Eng C 73:788–797
4. Naahidi S, Jafari M, Edalat F, Raymond K, Khademhosseini A, Chen P (2013) Biocompatibility of engineered nano-particles for drug delivery. J Controlled Release 166:182–194
5. Shagaldi BF, Compson J (2000) Wear and corrosion of sliding counterparts of stainless-steel hip screw plates. Inj Int J Care Injured 31:85–92
6. Bains PS, Grewal JS, Sidhu SS, Kaur S (2017) Wear between ring and traveler: a pin-on-disc mapping of various detonation gun sprayed coatings. Mater Today Proc 4(2):369–378
7. Manivasagam G, Dhinasekaran D, Rajamanickam A (2010) Biomedical implants: corrosion and its prevention—a review. Rec Pat Corros Sci 2:40–54
8. Fu T, Wen CS, Lub J, Zhou YM, Mac SG, Dong BH, Liu BG (2012) Sol-gel derived $TiO_2$ coating on plasma nitrided 316L stainless steel. Vacuum 86:1402–1407
9. Kaliaraj GS, Ramadoss A, Sundaram M, Balasubramanian S, Muthirulandi J (2014) Studies of calcium-precipitating oral bacterial adhesion on TiN, $TiO_2$ single layer, and TiN/$TiO_2$ multilayer-coated 316L SS. J Mater Sci 49:7172–7180
10. Sharifnabi A, Fathi MH, Yekta BE, Hossainalipour M (2014) The structural and bio-corrosion barrier performance of Mg-substituted fluorapatite coating on 316L stainless steel human body implant. Appl Surf Sci 288:331–340
11. Chang SH, Chen JZ, Hsiao SH, Lin GW (2014) Nanohardness, corrosion and protein adsorption properties of $CuAlO_2$ films deposited on 316L stainless steel for biomedical applications. Appl Surf Sci 289:455–461

12. Kathuria YP (2006) The potential of biocompatible metallic stents and preventing restenosis. Mater Sci Eng A 417:40–48

13. Wang KK, Kim BJ, Heo II, Jung SJ, Hwang JW, Kim YR (2017) Fabrication and characterization of antimicrobial surface-modified stainless steel for bio-application. Surf Coat Technol 310:256–262

14. Harun WSW, Asri RIM, Romlay FRM, Sharif S, Jan NHM, Tsumori F (2018) Surface characterisation and corrosion behaviour of oxide layer for SLMed-316L stainless steel. J Alloys Compd 748:1044–1052

15. Bekmurzayeva A, Duncanson WJ, Azevedo HS, Kanayeva D (2018) Surface modification of stainless steel for biomedical applications: revisiting a century-old material. Mater Sci Eng C 93:1073–1089. https://doi.org/10.1016/j.msec.2018.08.049

16. Asri RIM, Harun WSW, Samykano M, Lah NAC, Ghani SAC, Tarlochan F, Raza MR (2017) Corrosion and surface modification on biocompatible metals: a review. Mater Sci Eng C 77:1261–1274

17. Bains PS, Mahajan R, Sidhu SS, Kaur S (2019) Experimental investigation of abrasive assisted hybrid EDM of Ti-6Al-4V. J Micromanuf. https://doi.org/10.1177/2516598419833498

18. Li W, Xin Z, Ming-hui D, Hang Z, Hong-sen Z, Bin Z, Xun-qi L, Gao-yu W (2015) Surface modification of biomedical AISI 316L stainless steel with zirconium carbonitride coatings. Appl Surf Sci 340:113–119

19. Kumar AM, Rajendran N (2013) Electrochemical aspects and in vitro biocompatibility of polypyrrole/TiO$_2$ ceramic nanocomposite coatings on 316L SS for orthopedic implants. Ceram Int 39(5):5639–5650

20. Bains PS, Sidhu SS, Payal HS (2017) Investigation of magnetic field-assisted EDM of composites. Mater Manuf Process 33(6):670–675

21. Zain ZM, Ndaliman MB, Khan AA, Ali MY (2014) Improving micro-hardness of stainless steel through powder-mixed electric discharge machining. J Eng Sci 1–7

22. Marashi H, Sarhan AD, Hamdi M (2015) Employing Ti nano-powder dielectric to enhance surface characteristics in electrical discharge machining of AISI D2 steel. Appl Surf Sci 357:892–907

23. Talla G, Gangopadhayay S, Biswas CK (2016) State of art in powder-mixed electric discharge machining: a review. J Eng Manuf 231(14):2511–2526

24. Mahajan A, Sidhu SS (2017) Surface modification of metallic biomaterials for enhanced functionality: a review. Mater Technol 33(2):93–105

25. Prakash C, Kansal HK, Pabla BS, Puri S, Aggarwal A (2015) Electric discharge machining—a potential choice for surface modification of metallic implants for orthopedic applications: a review. J Eng Manuf 230(2):331–353

26. Prakash C, Uddin MS (2017) Surface modification of β-phase Ti implant by hydroaxyapatite mixed electric discharge machining to enhance the corrosion resistance and in-vitro bioactivity. Surf Coat Technol 326:134–145

27. Mazur A, Szczurek A, Checmanowski JG, Szczygiel B (2018) Corrosion resistance and bioactivity of SiO$_2$-Y$_2$O$_3$ coatings deposited on 316L steel. Surf Coat Technol 350:502–510

28. Kaliaraj GS, Vishwakarma V, Alagarsamy K, Kamalan Kirubaharan AM (2018) Biological and corrosion behavior of m-ZrO$_2$ and t-ZrO$_2$ coated 316L SS for potential biomedical applications. Ceram Int 44(12):14940–14946

29. Abbas NM, Solomon DG, Bahari MF (2007) A review on current research trends in electrical discharge machining (EDM). Int J Mach Tools Manuf 47(7–8):1214–1228

30. Abdul-Rani AM, Nanimina AM, Ginta TL, Razak MA (2017) Machined surface quality in nano aluminum mixed electrical discharge machining. Proc Manuf 7:510–517

31. Singh G, Bhui AS, Sidhu SS (2017) Influence of input parameters on MRR of AISI-316L using tungsten electrode machined by EDM. ISBN 978-5-398-01932-2:15-19

32. Sharma S, Kumar A, Beri N, Kumar D (2010) Effect of aluminum powder addition in dielectric during electric discharge machining of hastelloy on machining performance using reverse polarity. Int J Adv Eng Technol 1(3):13–24

33. Karamian E, Motamedi MRK, Khandan A, Soltani P, Maghsoudi S (2014) An in-vitro evaluation of novel NHA/zircon plasma coating on 316L stainless steel dental implant. Prog Nat Sci
Mater Int 24(2):150–156
34. Ramires PA, Romito A, Cosentino F, Milella E (2001) The influence of titania/hydroxyapatite
composite coatings on in vitro osteoblasts behavior. Biomaterials 22(12):1467–1474
35. Manam NS, Harun WSW, Shri DNA, Ghani SAC, Kurniawan T, Ismail MH, Ibrahim MHI
(2017) Study of corrosion in biocompatible metals for implants: a review. J Alloys Compd
70:698–715
36. Bains PS, Sidhu SS, Payal HS, Kaur S (2019) Magnetic field influence on surface modifications
in powder mixed EDM. Silicon 11(1):415–423
37. Long BT, Phan NH, Cuong N, Jatti VS (2016) Optimization of PMEDM process parameter for maximizing material removal rate by Taguchi's method. Int J Adv Manuf Technol
87(5–8):1929–1939
38. Sidhu SS, Bains PS (2017) Study of recast layer of particulate reinforced metal matrix composites machined by EDM. Mater Today Proc 4(2):3243–3251
39. Bains PS, Singh S, Sidhu SS, Kaur S, Ablyaz TR (2018) Investigation of surface properties of Al-SiC composites in hybrid electrical discharge machining. In: Futuristic composites.
Springer, Singapore, pp 181–196

# Chapter 9
# In Vitro and In Vivo Evaluation of 1-(3 Dimethylaminopropyl)-3-Ethyl Carbodiimide (EDC) Cross-Linked Gum Arabic–Gelatin Composite as an Ideal Porous Scaffold for Tissue Engineering

Boby T. Edwin, H. Dhanya, Prabha D. Nair and Moustapha Kassem

## 1 Introduction

Restoration of lost tissue and organs is one of the contemporary challenges of health care. Tissue engineering is an alternative to solve the deficiency of transplants and organs. It deals with the development of tissues over synthetic biocompatible polymers. This involves the fabrication of scaffolds and seeding of either patients' own cells or cell from other sources and differentiation to the required type of tissue. One of the key steps in tissue engineering is the development of scaffolds with optimum cell friendly properties. Ideally, a scaffold should be biocompatible with the controllable degradation rate, highly porous with interconnectivity, optimum mechanical strength and should be able to initiate chemical signals to guide tissue growth.

The success of regenerative medicine lies mainly in the development of materials having defined positive interaction with cells. Porous matrices with predefined structural and mechanical properties—hydrogels, sponges and fibrous meshes have been widely used as three-dimensional supports for adhesion of the cell, proliferation and ECM formation [1–6]. Natural polymer-based porous matrices are synthesized from biopolymers, like cellulose [7] hyaluronic acid [8] collagen [9], chitosan [10] and

B. T. Edwin (✉)
TKM College of Arts and Science, Kollam, India
e-mail: bobytedwin2003@gmail.com

H. Dhanya
Cashew Export Promotion Council, Kollam, India

P. D. Nair
Regenerative Medicine and Tissue Engineering, SCTIMST, Thiruvananthapuram, India

M. Kassem
Medical Biotechnology Center, Southern Denmark University, Odense, Denmark

© Springer Nature Singapore Pte Ltd. 2019                                      131
P. S. Bains et al. (eds.), *Biomaterials in Orthopaedics and Bone Regeneration*,
Materials Horizons: From Nature to Nanomaterials,
https://doi.org/10.1007/978-981-13-9977-0_9

silk fibroin [11], for tissue engineering (TE) as they provide a suitable and structural cellular environment. Gum Arabic and *Gelatin* are excellent natural raw materials for the design of porous matrices with added advantages in terms of biocompatibility, chemistry versatility and controlled degradability because of their intrinsic characteristics [12]. Though self-cross-linking scaffold using gum Arabic and gelatin is developed using Schiff's base method, the scaffold was found to be non-attaching and found suitable for managing the non-attached type of cells like spheroids [13]. Hence, the requirement of a slow-degrading scaffold with optimum cell adhesion was a challenge.

Gelatin, a porous denatured collagen scaffold, used as a scaffolding structure for tissue engineering. Gelatin is a collagen-derived product having natural origin; it is an attractive biopolymer candidate for fabricating tissue engineering scaffold. However, the major limitation with type I collagen scaffolds and type II collagen scaffolds was their inability to preserve the chondrocyte phenotype. The lower mechanical strength was a problem with gelatin scaffolds [14]. The goal of this study was to fabricate a scaffold with better mechanical strength and slow degrading by adding gum Arabic and to assess the functional properties of gelatin–gum Arabic scaffold for its future applications. The present work reports the first attempt to develop a cross-linked, porous composite scaffold from gelatin and gum Arabic with optimum cell adhesion and controlled degradation using EDC

## 2   Materials and Methods

Gelatin (G-2625), derived from porcine skin, Gum Arabic (G-9752), derived from acacia tree were purchased from Sigma-Aldrich (India) and 1-(3 Dimethylaminopropyl)-3-Ethyl Carbodiimide (EDC) and all other chemicals purchased from SRL Chemical India.

### 2.1   Fabrication of the Gelatin–Gum Arabic Scaffold

The scaffold fabrication was carried out by foaming and freeze-drying method [15]. Twenty wt% of gelatin was prepared by dissolving gelatin in water in lukewarm condition, and two wt% of gum Arabic was prepared in water. Gelatin was stirred at 1800 rpm for 10 min. One by third of the volume of gum Arabic solution was added to the gelatin and continued stirring for 25 min. The thick foam formed poured into plastic vials of 2.5 cm diameter and freeze-dried for 67 h. The formed porous matrix after removing the upper and lower portion is cut into small pieces and cross-linked with EDC in methanol for 24 h. The concentration of EDC was 1 mg/40 mg of the scaffold.

## 2.2   Morphological Characterization (SEM)

Scanning electron microscopy (SEM) images were captured using 30 kV environmentally scanning electron microscope (ESEM Quanta 200, Germany). SEM images were captured on dry scaffolds, scaffolds wet in PBS and also with cells seeded on them to observe the porosity of the cell in wet and dry condition and adhesion of the cells to the scaffold.

## 2.3   Micro-CT and Confocal Imaging

3D structure of the scaffold was quantified by micro-computed tomography and visualized by imaging. Scaffolds were scanned (SCANCO Medical) at a resolution of $2048 \times 2048$ pixels (12 $\mu$m thick slices). 3D images of the scaffolds were generated after reconstruction. The 3D distribution and growth of cells on scaffolds were captured with the confocal microscope (Carl Zeiss) after staining with LIVE/DEAD™ Cell Imaging Kit (R37601). Epiplain-Neofluar 20x/0.50 HD objective was used with the laser beam of 514 nm wavelength excitation. Image stacks of $512 \times 512 \times 30$ with a scaling of $0.88 \ \mu M \times 0.88 \ \mu M \times 10 \ \mu M$ were made. The Carl Zeiss LSM image examiner was used for image processing.

## 2.4   Mechanical Strength, Swelling and Degradation Analysis

Mechanical strength was assessed by compression tests using a universal testing machine (Instron series IX) with a load cell of 100 N. Scaffold samples with dimensions 10 mm length and 5 mm diameter were used for analysis. The water uptake or swelling analysis was performed on five specimens of the scaffold (diameter 5 mm, thickness 2 mm and 40 mg). The scaffolds immersed in PBS (PH 7) after recording the dry weight, wet weights recorded at an interval of 1 h and the swelling percentage was calculated. To measure the hydrolytic degradation of the scaffold, dry scaffolds ($0.2 \times 0.5$ cm) were weighed and immersed in 10 mL phosphate-buffered saline (PBS, pH = 7.4) at 37 °C in a water bath. Samples were taken at intervals and weighed after drying in vacuum. The process continued for one week. The weight remaining was calculated as—weight remaining (%) $= 100 \times w_2/w_1$, where $w_1$ and $w_2$ are the weights of the scaffold before and after degradation.

## 2.5   Cell Culture and Scaffold Seeding

Ovine bone marrow stem cells (OVBMSC) were isolated from cadaver born marrow by centrifugation and plastic adherence. The cells were cultured in DMEM containing 10% fetal bovine serum (B D bioscience) and 1% antibiotic solution (100 $\mu$g/mL penicillin, 100 $\mu$g/mL streptomycin). The third passage cells were used in all biological studies. The cells were characterized by staining with the vimentin antibody. Static surface seeding was performed for seeding cells in the scaffolds. The scaffolds were pre-soaked for 5 h in media in a bio-safety cabinet in a six-well plate. After soaking, excess media around the scaffold was aspirated. Concentrated cell suspension ($1 \times 10^6$ cells/10 $\mu$L) was injected into the center of the scaffold using a 25-gauge pipette, as described in other reports [16–18]. The scaffold was allowed to incubate for 3 h which was the optimum time required for cell attachment, as observed in control wells. After incubation, the scaffold was submersed in DMEM with serum and antibiotics. Scaffolds were retained with media change at an interval of 48 h for three weeks.

## 2.6   Implantation of the Scaffold in Mice for In Vivo Cytotoxicity and Apoptosis Assay

The eyes of the NOD-scid mice were protected with the application of the ophthalmic liquid gel, and the mice were anesthetized using 2% isoflurane USP-PPC. Hairs in the back of mice were removed by shaving, and the underlying skin cleaned and sterilized. Animal was hydrated with a subcutaneous injection of 1 ml of 0.9% sodium chloride solution. Optimum sterility measures were practiced for survival surgeries. Incisions of 8 mm were cut on the dorsal section of each mouse for implanting the scaffolds. Two GELGA scaffold samples were separately implanted into each mouse. The incisions sutured and transdermal bupivacaine 2% was topically applied to the surgery sites for stoping infections. The animals were then keenly observed for the following three days.

## 2.7   Scaffold Resections and Histology

At 7, 14 and 21 days after scaffold implantation, the mice were euthanized using $CO_2$ inhalation. Resected dorsal skin was immersed in PBS solution instantly. The skin sections containing GELGA scaffolds were then photographed, cut and fixed in 10% formalin for at least 48 h. The samples were then kept in 70% ethanol before being embedded in paraffin by the Histology Facility of the medical biotechnology center Odense. The serial 10 $\mu$m thick sections were cut 1 mm inside the GELGA scaffold and stained with hematoxylin–eosin (H&E). For immunocytochemistry analysis,

(a)                          (b)                          (c)

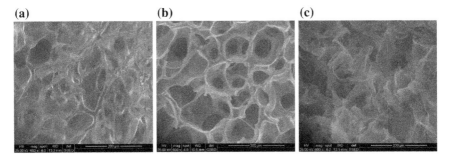

**Fig. 1** SEM micrographs of GELGA. **a** Dry GELGA showing interconnected pores. **b** GELGA soaked in the media revealing the interconnectivity. **c** One month PBS soaked GELGA revealing the partial degradation

heat-induced epitope retrieval was used [110 °C for 12 min with citrate buffer (pH 6.0)]. The apoptotic properties were also checked by staining with the caspase-3 antibody. The sections were observed under the Olympus Microscope loaded with the 40x objective and analyzed using the Pannoramic Viewer.

# 3  Results

## 3.1  Fabrication of the Scaffold, Physical and Biological Characterization

### 3.1.1  Scanning Electron Microscopy

The morphology of the GELGA was analyzed by SEM. SEM images of the porous GELGA composite show open porosity with pore sizes ranging from 40 to 100 μm (Fig. 1a). This type of porosity is considered to be suitable for growing cells. The scaffolds in a soaked condition with a slight reduction in pore size from 40 to 90 μm were observed (Fig. 1b). The scaffold soaked in PBS after one month revealed partial degradation under SEM. (Fig. 1c). Cells seeded on GELGA demonstrated good growth with filopodia formation and massive deposition of the extracellular matrix one month after seeding. The walls of the scaffolds have been degraded and replaced with extracellular matrix from OVBMSC showing re-alignment of the scaffold due to biodegradation.

### 3.1.2  Micro-computed Tomography and 3D Imaging of the Scaffold

The porosity, pore size and pore percentage of the scaffold were quantitatively measured by micro-CT. The porosity percent was around 71.83% for dry samples. The pore size extended from 12 to 156 μM. The percentage pore volume was measured. The major pore sizes include 84 μM (24%), 96 84 μM (15%), 72 μM (19%) and 60 μM (17%). In wet samples, the pore size extended from 12 to 92 μM. The major pore was 92 μM (13%). The total pore volume in the dry scaffold was 6.1935 mm$^3$, while the material volume was 2.2428 mm$^3$ and the total volume 8.6215 mm$^3$. The average wall thickness and pore size in the dry scaffold were 34 and 79 μM, respectively. In wet GELGA, the total pore volume was 4.513 mm$^3$, while the material volume was 32.2428 mm$^3$ and the total volume 37.4208 mm$^3$. The average wall thickness and pore size in the dry scaffold were 59 and 42 μM, respectively (Fig. 2).

Seven days after BMSC seeding, the constructs were stained with live dead assay and images were captured on a confocal microscope (Fig. 3). The stack of images revealed the presence of live, attached growing cells on the pore walls of the scaffold. The cells also revealed spreaded morphology.

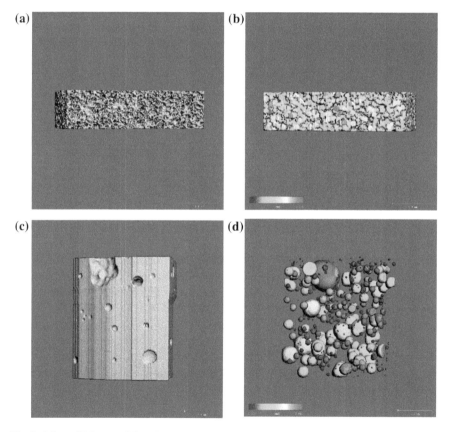

**Fig. 2** Micro-CT image of GELGA showing wall and pore distribution. **a, b** Dry. **c, d** Soaked

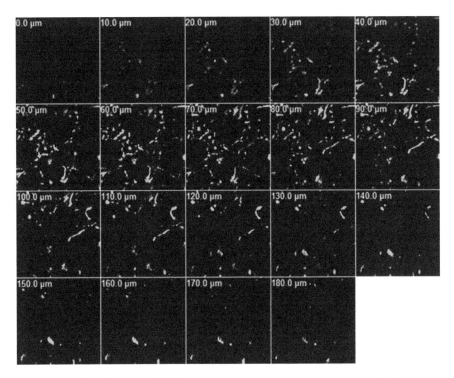

**Fig. 3** The confocal image of GELGA revealing the attachment and progression of cells in stacks

### 3.1.3 Compressive Strength, Swelling and Biodegradation

To examine the mechanical property of the scaffold, the compressive strength of the GELGA scaffolds were measured in dry condition (Fig. 4). The compressive

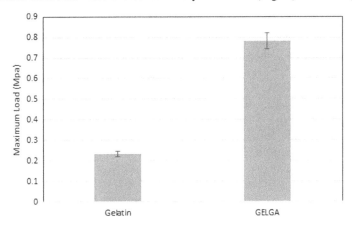

**Fig. 4** Compressive strength of gelatin scaffold and GELGA

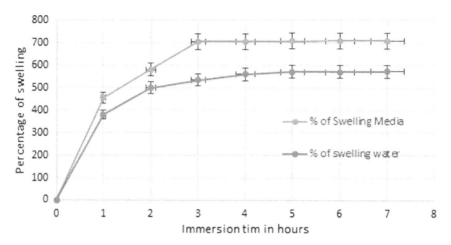

**Fig. 5** Variation in swelling percentage of GELGA in water and media

strength of GELGA was found to be 0.783 ± 15 MPa. In comparison to gelatin, the mechanical strength was increased approximately fourfold. The scaffold attained maximum swelling with in 5 h when soaked in water as well as in DMEM. The swelling studies revealed that final swelling percentages of GELGA varied from 400 to 700% (Fig. 5). The volume of scaffolds increased during swelling. No significant biodegradation was observed in the early phase of degradation upto one week. Extend of biodegradation slightly increased with the increase in time from seven days. After 21 days only 40% degradation was observed.

### 3.1.4 FTIR and Thermal Analysis

The FTIR spectrum of gum Arabic, gelatin and GELGA is shown in Fig. 6. Characteristic peaks of Gum Arabic were clearly observed at 3276 cm$^{-1}$, which were mainly due to the OH-stretching vibrations. The broadness of the peak is due to the presence of many OH groups. –CH$_2$-stretching peaks were observed at 2929 cm$^{-1}$, and the wagging CH$_2$ peaks were observed at 1598 cm$^{-1}$. The peaks at 1413 and 1016 cm$^{-1}$ indicate the C–O–C plane bending vibration.

The thermal characteristics of GELGA were analyzed by TGA (Fig. 7). Water will be a part of the composite, and the endothermic peak initiated at 80 °C confirms the energy absorption during water evaporation. The fall of the peak starting from 100 °C is due to the moisture loss. In the case of gelatin, the spectrum was separated into three major regions. The first region in the range of 3600–2700 attributes amide linkage and stretching vibrations. The second region was between 1630 and 1500 corresponding to the coil structure of gelatin. The peak at 1522 also indicates the presence of H-bond coupled with NH$_2$ linkage. The peak at 1437 corresponds to the ester linkage. The spectrum of GELGA is almost similar to gelatin. It can be assigned that every amide and ester linkages were retained.

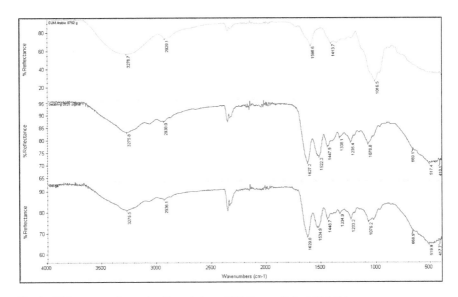

**Fig. 6**  FTIR spectra of gum Arabic, gelatin and EDC cross-linked GELGA

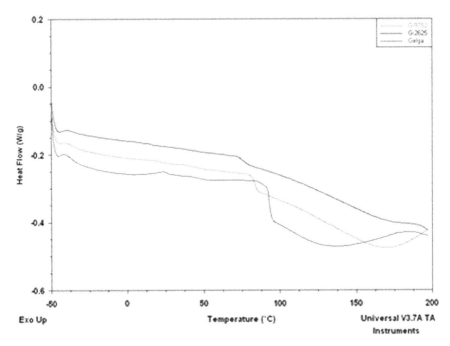

**Fig. 7**  The TGA curve of gelatin, gum Arabic and EDC cross-linked GELGA

### 3.1.5   Adhesion and Growth of BMSCs on GELGA

After 12 h, the cells seeded on the scaffold started to stretch to attain the spindle-shaped morphology. After 14 days, BMSCs reached about 60% confluence under the confocal image (Fig. 8A). The cells were found to be distributed on the pore walls of the scaffold. The SEM image of a one month construct revealed the presence of a highly oriented cell layer on the scaffold with numerous mass of extracellular matrix (Fig. 8B).

### 3.1.6   In Vivo Biocompatibility of GELGA

To examine the biocompatibility of GELGA in vivo, scaffolds were subcutaneously implanted into mice. The mice were sacrificed to excise the implants on seven, 14 and 21 days. Histological images showed a gradual degradation of GELGA over 21 days of implantation with migration of cells and tissue formation. H&E staining revealed the absence of inflammatory reactions on the sides of the scaffold (Fig. 8C). No cytotoxic developments are observed in the slides, and the scaffold is found to be cell friendly. The staining for caspase-3 was found to be negative revealing non-apoptotic nature of the scaffold.

## 4   Discussion

As scaffolds for tissue engineering, especially for bone and cartilage, biodegradable scaffold GELGA was synthesized combining gelatin and gum Arabic. The gelatin and gum Arabic were selected for optimizing the mechanical property of the scaffold, one soft material gelatin and one hard material gum Arabic both being biological in origin. EDC in ethanol was used as a cross linker. The molecular weights of polymers used were high, glass transition temperatures were low and hence the scaffolds were slightly flexible [19]. As gelatin in aqueous media was self-foamable on stirring, foaming and freeze-drying method was used as a method to make scaffolds with an open interconnected pore with extended pore walls for maximum cell attachment and migrated growth [20].

As both the biochemicals selected are soluble in water, water was the medium for scaffold fabrication. A slight temperature was used initially to facilitate solubilization of gelatin. The polymer solution concentration is expected to affect pore morphology and porosity. When the concentration of the polymer solution is high, the viscosity of the solution increases and forms scaffolds with smaller pores and better mechanical strength. As both the biopolymers were hygroscopic and viscous, the foam formation was relatively quick. Due to viscosity, the rotation at 2500 rpm formed a thick scaffold with pores. The optimum concentration of the polymer solution was determined by trial and error method.

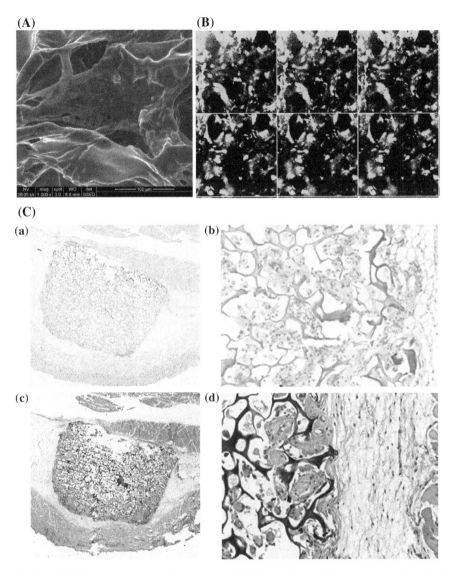

**Fig. 8** **A** ESEM image of one month construct of GELGA seeded with the ovbmscs. **B**. Confocal image of live dead assay of 21 day construct seeded with the ovbmscs. **C** Hematoxylin Eosin and caspase staining of the scaffold implanted in NOD-scid mice. **a** Caspase-3 staining. The absence of brown spots indicates the non-apoptotic nature of the scaffold. **b** A portion of the caspase stained implant zoomed. **c** HE staining of the construct showing cell migration and growth. **d** A portion of the HE stained implant zoomed

A scaffold suitable for tissue engineering needs the following features: a suitable 3D structure for cell attachment, growth and nutrient transport and optimal pore size to prevent loss of cell from the scaffolds. Under SEM and micro-CT, we found that the GELGA scaffolds had random porous microstructures. Porosity allows maximum interaction of the scaffold with the cells [21, 22]. Cell seeding efficiency is determined by pore size [23, 24]. In tissue engineering scaffolds, large pores prevent cell attachment due to a reduced surface area while small pores may result in pore occlusion by the cells leading to lesser migration and growth. The pore size above 50 mM and below 150 mM is found to be ideal for tissue engineering. The maximum pore size of the scaffold was 150 mm while the maximum percentage of the pore was 90 mM. The pore was ideal for cell growth and development as proved by extensive growth of the cells observed under SEM analysis, confocal imaging and implantation.

The in vitro degradation studies demonstrated that degradation of the scaffold was in a controlled model at rates attractive for tissue engineering applications especially bone and cartilage tissue engineering. Due to the presence of hygroscopic gelatin, the scaffold becomes swollen with in 4–5 h. Thus, less pre-soaking period is enough for GELGA before cell seeding.

The scaffold was further characterized for the properties using FTIR and TGA. From the FTIR spectra, we can ascertain that in the presence of cross linker, gum Arabic structure was intercalated in gelatin, which results in the formation of a polymer alloy. Also, this intercalation of gum Arabic (GA) to gelatin can result in the absence of characteristic peaks of gum Arabic. Comparison can be done only on the basis of peak intensity, and it has been reported that [25]. Due to the above factors, we claim that gelatin is modified by GA and thus GELGA is a stable composite. TGA studies also revealed that the mechanical strength of GELGA has increased due to the addition of gum Arabic in a trace amount and thermal stability has been attained. The addition of gum Arabic enhanced the thermal stability of GELGA. Although GELGA revealed some sharp decrease in the mechanical strength, the final product showed enhanced thermal stability, at least near to the parent compounds. The initial decrease in spectral lines is mainly due to water loss. It was mainly observed at the temp 100 °C. After moisture loss, no drastic variation was observed in three peaks. But GELGA showed some extra stability with can be observed from the thermogram.

A prime concern in developing a tissue engineering biodegradable scaffold for implantation is the absence of cytotoxicity, the absence of cytotoxic or carcinogenic degradation products or the release of residual solvent or cross linker used during synthesis. Biomaterial gelatin is extensively used in the medical field and is a non-toxic component. Gum Arabic though it is a plant-derived, reports are available regarding its toxic nature. EDC cross linker is a toxic compound to the cells [26]. The scaffold fabricated revealed no toxicity probably due to extensive washing procedure and may be due to the alloy effect [4, 27]. No matter in vivo or in vitro, cells and tissues will first contact with the surface of material. So, it is very important for the cells to adhere to the material surface as the first step. Adhesion properties of different materials can largely affect the proliferation and differentiation of the cells [28]. Excellent proliferation, adhesion and spread of BMSCs on GELGA in vitro were analyzed by SEM and confocal imaging. By analyzing the scaffolds using confocal microscopy using live dead stains, we qualitatively observed a larger cell penetration and even dispersion in GELGA on 14th-day and 21st-day cell seeded scaffolds. The porosity and the hydrophilic nature of gelatin in the scaffold may be the reason for this promising cell dispersion. On confocal stacks, cells were found to be more concentrated on the mid portion of the scaffold. This indicates that the porosity is optimum for the scaffold to penetrate to the inner side. Also, the morphology and spreading of the cells indicated the non-cytotoxic nature of the scaffold.

Meanwhile, SEM results showed that the three-dimensional porous structure of GELGA favored secretion and infiltration of nutrients and metabolites. The SEM image clearly shows extensive deposition of the extracellular matrix. Thus, the material was found to be cytocompatible.

## 5  Conclusion

In vivo implantation studies followed by the excision of the scaffold and HE and caspase-3 staining also revealed the cytofriendly and non-apoptotic nature of the scaffold. The absence of brown coloration in the caspase staining shows non-expression of caspase-3, the effector molecule responsible for programmed cell death. The in vitro and in vivo experiments using GELGA revealed non-toxicity, cytocompatibility, controlled degradation and optimum swelling parameters optimum for a tissue engineering scaffold. Hence, considering the mechanical properties, the scaffold will be ideal for tissue engineering especially for bone and cartilage.

**Acknowledgements** The authors are thankful to The DST SERB project facility SR/SO/HS/81/2010 and the FIST facility DST/FIST-SR/FST/College-213/2014 of TKM College of Arts and Science.
**Ethical Approval** All applicable international, national and/or institutional guidelines for the care and use of animals were followed.

## References

1. Keane TJ, Badylak SF (2014) Biomaterials for tissue engineering applications. Semin Pediatr Surg. https://doi.org/10.1053/j.sempedsurg.2014.06.010
2. Loh QL, Choong C (2013) Three-dimensional scaffolds for tissue engineering applications: role of porosity and pore size. Tissue Eng Part B Rev. https://doi.org/10.1089/ten.teb.2012.0437
3. Ma PX (2004) Scaffolds for tissue fabrication. Mater Today. https://doi.org/10.1016/S1369-7021(04)00233-0
4. O'Brien FJ (2011) Biomaterials and scaffolds for tissue engineering. Mater Today 14(3):88–95. http://dx.doi.org/10.1016/S1369-7021(11)70058-X
5. Tan H, Marra KG (2010) Injectable, biodegradable hydrogels for tissue engineering applications. Materials. https://doi.org/10.3390/ma3031746

6. Guan J, Fujimoto KL, Sacks MS, Wagner WR (2005) Preparation and characterization of highly porous, biodegradable polyurethane scaffolds for soft tissue applications. Biomaterials 26(18):3961–3971
7. Müller FA, Müller L, Hofmann I, Greil P, Wenzel MM, Staudenmaier R (2006) Cellulose-based scaffold materials for cartilage tissue engineering. Biomaterials. https://doi.org/10.1016/j.biomaterials.2006.02.031
8. Collins MN, Birkinshaw C (2013) Hyaluronic acid based scaffolds for tissue engineering—a review. Carbohydr Polym 92(2):1262–1279. http://dx.doi.org/10.1016/j.carbpol.2012.10.028
9. Yannas IV, Tzeranis DS, Harley BA, So PTC (2010) Biologically active collagen-based scaffolds: advances in processing and characterization. Philos Trans Math Phys Eng Sci 368(1917):2123–2139. Retrieved from http://www.jstor.org/stable/25663359
10. Dhandayuthapani B, Krishnan UM, Sethuraman S (2010) Fabrication and characterization of chitosan-gelatin blend nanofibers for skin tissue engineering. J Biomed Mater Res—Part B Appl Biomater 94(1):264–272. https://doi.org/10.1002/jbm.b.31651
11. Yao D, Liu H, Fan Y (2016) Silk scaffolds for musculoskeletal tissue engineering. Exp Biol Med (Maywood, N.J.) 241(3):238–245. https://doi.org/10.1177/1535370215606994
12. Tsai RY, Kuo TY, Hung SC, Lin CM, Hsien TY, Wang DM, Hsieh HJ (2015) Use of gum arabic to improve the fabrication of chitosan-gelatin-based nanofibers for tissue engineering. Carbohyd Polym 115:525–532. https://doi.org/10.1016/j.carbpol.2014.08.108
13. Sarika PR, Cinthya K, Jayakrishnan A, Anilkumar PR, James NR (2014) Modified gum arabic cross-linked gelatin scaffold for biomedical applications. Mater Sci Eng C 43:272–279. https://doi.org/10.1016/j.msec.2014.06.042
14. Xing Q, Yates K, Vogt C, Qian Z, Frost MC, Zhao F (2014) Increasing mechanical strength of gelatin hydrogels by divalent metal ion removal. Sci Rep 4:4706. https://doi.org/10.1038/srep04706
15. Lu T, Li Y, Chen T (2013) Techniques for fabrication and construction of three-dimensional scaffolds for tissue engineering. Int J Nanomed 8:337–350. https://doi.org/10.2147/IJN.S38635
16. Hofmann A, Konrad L, Gotzen L, Printz H, Ramaswamy A, Hofmann C (2003) Bioengineered human bone tissue using autogenous osteoblasts cultured on different biomatrices. J Biomed Mater Res. Part A 67(1):191–199. https://doi.org/10.1002/jbm.a.10594
17. Honda MJ, Yada T, Ueda M, Kimata K (2004) Cartilage formation by serial passaged cultured chondrocytes in a new scaffold: hybrid 75:25 poly(L-lactide-ε-caprolactone) sponge. J Oral Maxillofac Surg 62(12):1510–1516. https://doi.org/10.1016/j.joms.2003.12.042
18. Thevenot P, Nair A, Dey J, Yang J, Tang L (2008) Method to analyze three-dimensional cell distribution and infiltration in degradable scaffolds. Tissue Eng Part C Methods 14(4):319–331. https://doi.org/10.1089/ten.tec.2008.0221
19. Fang Y, Li L, Vreeker R, Yao X, Wang J, Ma Q, Phillips GO (2011) Rehydration of dried alginate gel beads: effect of the presence of gelatin and gum arabic. Carbohyd Polym 86(3):1145–1150
20. Lv Q, Feng QL (2006) Preparation of 3-D regenerated fibroin scaffolds with freeze drying method and freeze drying/foaming technique. J Mater Sci Mater Med. https://doi.org/10.1007/s10856-006-0610-z
21. Tsuruga E, Takita H, Itoh H, Wakisaka Y, Kuboki Y (1997) Pore size of porous hydroxyapatite as the cell-substratum controls BMP-induced osteogenesis. J Biochem 121(2):317–324
22. Lyons F, Partap S, O'Brien FJ (2008) Part 1: scaffolds and surfaces. Technol Health Care Official J Eur Soc Eng Med 16(4):305–317
23. Drury JL, Mooney DJ (2003) Hydrogels for tissue engineering: scaffold design variables and applications. Biomaterials. https://doi.org/10.1016/S0142-9612(03)00340-5
24. Karageorgiou V, Kaplan D (2005) Porosity of 3D biomaterial scaffolds and osteogenesis. Biomaterials. 26(27):5474–5491
25. Olsson AM, Salmén L (2004) The association of water to cellulose and hemicellulose in paper examined by FTIR spectroscopy. Carbohydr Res. https://doi.org/10.1016/j.carres.2004.01.005
26. Prata AS, Zanin MHA, Ré MI, Grosso CRF (2008) Release properties of chemical and enzymatic crosslinked gelatin-gum Arabic microparticles containing a fluorescent probe plus vetiver essential oil. Colloids Surf B Biointerfaces. https://doi.org/10.1016/j.colsurfb.2008.08.014

27. Hao Y, Xu P, He C, Yang X, Huang M, Xing J, Chen J (2011) Impact of carbondiimide crosslinker used for magnetic carbon nanotube mediated GFP plasmid delivery. Nanotechnology. https://doi.org/10.1088/0957-4484/22/28/285103
28. Fayyazbakhsh F, Solati-Hashjin M, Keshtkar A, Shokrgozar MA, Dehghan MM, Larijani B (2017) Novel layered double hydroxides-hydroxyapatite/gelatin bone tissue engineering scaffolds: fabrication, characterization, and in vivo study. Mater Sci Eng C. https://doi.org/10.1016/j.msec.2017.02.172

# Chapter 10
# Extraction of Hydroxyapatite from Bovine Bone for Sustainable Development

Emon Barua, Payel Deb, Sumit Das Lala and Ashish B. Deoghare

## 1 Introduction

Tissue engineering is an interdisciplinary branch of science that combines the principle of engineering and basic science for biomedical applications. The primary aim of tissue engineering (TE) is to fabricate biological alternatives that help in restoring, maintaining, and improving tissue functions [1]. Tissue damage is a major health concern these days owing to accidents or other congenital diseases. A number of techniques have been developed to repair or replace these damage tissues which include the use of bone grafts like allograft and autograft [2]. Allograft is a bone or tissue that is transplanted from one person to another. The major drawback of this tissue replacement is that there is always a risk of rejection or tissue growth failure, donor shortage, and the process is highly expensive. However, autograft is a bone or tissue that is transferred from one part of the body to the other in the same person's body. But the major limitations of this tissue replacement are post-operation pains and longer time of recovery from the second surgery [3]. This has lead to the development of artificial bone material such as tissue engineered scaffold which can repair damage tissue and act as a template for tissue growth.

E. Barua · P. Deb (✉) · S. Das Lala · A. B. Deoghare
Department of Mechanical Engineering, National Institute of Technology,
Silchar 788010, Assam, India
e-mail: payeldebmech13@gmail.com

E. Barua
e-mail: imon18enator@gmail.com

S. Das Lala
e-mail: sumitdaslala@gmail.com

A. B. Deoghare
e-mail: ashishdeoghare@gmail.com

© Springer Nature Singapore Pte Ltd. 2019                                    147
P. S. Bains et al. (eds.), *Biomaterials in Orthopaedics and Bone Regeneration*,
Materials Horizons: From Nature to Nanomaterials,
https://doi.org/10.1007/978-981-13-9977-0_10

A number of biomaterials have emerged, in recent years, [4] for the development of artificial bone materials. Among them, hydroxyapatite (HAp) has gained huge attention from researchers and scientists working in the field of medical science. HAp is a calcium-phosphate-based ceramic compound having excellent biocompatibility, easily tunable physical–chemical properties, non-toxicity, excellent storage stability, and inertia to microbial degradation [5]. It has similar properties to that of human bone and is finding the utmost importance in the field of orthopedics and dental implants [6]. HAp can be synthesized either synthetically from different reagents and chemicals or can be extracted from various natural resources.

Although synthetic HAp shows the exact stoichiometric ratio of calcium and phosphate similar to human bone, naturally synthesized HAp shows better cell adhesion and proliferation compared to synthetic HAp [7]. HAp extracted from natural resources such as fish scale, chicken bone, and eggshell show good biocompatibility, presence of desired minerals attached to it and non-toxicity which make it suitable for medical applications. Ho et al. [8] developed hydroxyapatite from eggshells by solid-state reactions between eggshell powders and dicalcium phosphate dihydrate ($CaHPO_42H_2O$, DCPD) as well as calcium pyrophosphate ($Ca_2P_2O_7$) which shows traces of different mineral components like Na, Mg, and Sr in the synthesized HAp. Panda et al. [9] extracted HAp from *Labeo rohita* and *Catla catla* fish scale and compared it with the HAp prepared from simulated body fluid (SBF) using wet precipitation method. The biocompatibility test confirms that HAp obtained from fish scales are physicochemically and biologically equivalent to the HAp chemically synthesized from SBF. Rajesh et al. [10] extracted hydroxyapatite from chicken bone bio-waste. The researchers concluded that the temperature range from 600 to 1000 °C shows the removal of organic matters from chicken bone leaving behind HAp ceramic. Therefore, chicken bone bio-waste can be a suitable bio-resource for extraction of HAp for bone tissue engineering. Rana et al. [11] extracted HAp from bovine bones by annealing the bones at high temperature. Similarly, Khoo et al. [12] derived crystalline HAp by the calcination of bovine femur at different temperatures and found that calcination temperature above 700 °C is suitable for deriving HAp for bone tissue engineering applications. However, the effect of deproteinization of bones prior to calcination was not considered in their study which has a significant effect on the properties of the synthesized HAp. Moreover, the percentage yield of HAp extracted from bovine bone was not explored by the researchers.

In the present study, an attempt is made to extract HAp from bovine bone bio-waste using thermal decomposition method. The bones are degreased and deproteinized by immersing in acetone and acid solution simultaneously one after another before heat treatment. The developed HAp is found to have improved crystallographic properties compared to the HAp synthesized by Rana et al. [11] and Khoo et al. [12], which can be a potential value-added product from bone bio-waste for biomedical applications.

## 2   Materials and Methods

### 2.1   Thermal Decomposition of Bovine Bone for HAp Extraction

Bovine bones were collected and washed thoroughly with tap water to remove debris, foreign materials, and impurities adhered on its surface. The washed bones were at first degreased by dipping in acetone solution in room temperature for 24 h. Since acetone is a non-polar solution, it can dissolve non-polar fat completely thereby degreasing bone materials. The bones were washed thoroughly with distilled water and were crushed into smaller pieces. Deproteinization of the crushed bones was carried out by immersing the bones in 1(N) HCl solution for 24 h. The amino acids that were building units of proteins get dissolved in HCl solution completely thereby deproteinizing the bones. The bones were then dried in room temperature for 24 h. Thereafter, they were oven dried in a hot air oven at 70 °C for 48 h. Finally, the bones were calcined at 1000 °C for 3 h using muffle furnace to obtain HAp ceramic. The detailed pictorial representation adopted for the extraction of HAp from bovine bones is depicted in Fig. 1.

### 2.2   Confirmation Test for HAp Formation

The formation of HAp was tested using XRD and FTIR analysis.

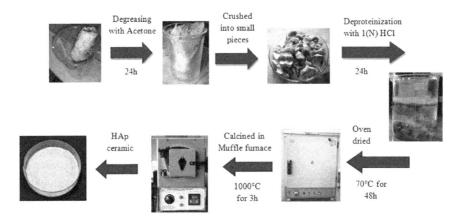

**Fig. 1**  Process followed for the calcination of bovine bone to extract HAp

### 2.2.1 X-Ray Diffraction (XRD)

It is an analytical technique adopted for the identification of phase in a crystalline material and provides crystal dimensions. The basic principle of XRD is interference occurring between the monochromatic X-rays and a crystalline material as shown in Fig. 2. The X-rays originate from cathode tube and are directed toward the finely powdered samples placed on the sample holder. The X-ray when interacts with the sample produces diffracted rays that are counted, detected, and processed.

The crystalline structure and the phase composition of the calcined powder were determined using X-ray diffractometer of PANalytical Model-X'Pert Pro. The scanning angle range was varied from 20° to 60°. The working voltage and current were set to 40 kV and 20 mA with a counting time of 2 s/step.

Phase identification was done by comparing the experimental XRD pattern to standards complied by the International Centre for Diffraction Data (ICDD) using card no. 09-0432 for hexagonal HAp structure. ICDD is a non-profit scientific organization solely devoted to collection, publishing, and distribution of diffraction pattern for the identification of materials. Patterns may be obtained through experimentation or determined based on the computation on crystal structure and Bragg's law.

The crystallite size of the HAp particles was calculated from Scherer equation as shown in Eq. 1 [13].

$$\tau = \frac{K\lambda}{\beta \cos \theta} \tag{1}$$

where,

$\tau$  Mean size (nm)
$K$  Dimensionless shape factor 0.9
$\lambda$  Wavelength (nm)
$\beta$  Line broadening at half the maximum intensity (FWHM) (degree)
$\theta$  Bragg angle (degree)

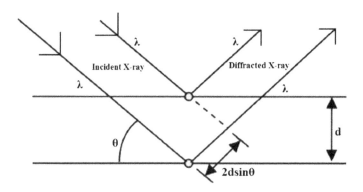

**Fig. 2**  Basic principle of XRD

Each atom in a crystal is represented as a single point that is joined to the other atom by line. The lattice is sub-divided into similar blocks. The interconnecting edges of unit cells define a set of crystallographic axes. The Miller indices are obtained by the intersection of the axes and plane. The reverse of the above intercepts is computed to obtain Miller indices (hkl). These hkl planes obtained from the XRD plots represent the orientation of atoms. Five high-intensity peaks corresponding to hkl planes (002), (211), (112), (300), and (222) were considered to calculate average crystallite size of HAp particle.

The percentage crystallinity was calculated from Eq. 2 as reported by Landi et al. [14] and Pang and Bao [15]. It gives the measure of degree of crystallinity of HAp particle.

$$X_c = 1 - \frac{V_{112/300}}{I_{300}} \tag{2}$$

where,

$X_c$      Degree of crystallinity
$I_{300}$      Intensity of 300 plane
$V_{112/300}$  Intensity of the valley between 300 and 112 planes.

### 2.2.2  Fourier-Transform Infrared Spectroscopy (FTIR)

It is a technique to obtain infrared spectrum of a material. The spectrum is generated by vibrational motion of the bonds present in molecules. The vibration can be either stretching or bending depending upon the nature of molecule. Stretching vibration occurs due to the change in inter-atomic distance along bond axis and bending vibration occurs due to the change in angle between two bonds in a molecule. The spectrometer accumulates these high-resolution vibrational data over wide range of spectra. Thus, it detects the presence of organic, inorganic, and biological compounds in a material.

FTIR analysis was conducted on Spectrum one FTIR spectrometer with a scan range varying from MIR 400–4000 cm$^{-1}$ and resolution of 1.0 cm$^{-1}$. The analysis was conducted on FTIR system of Make-Bruker and Model-3000 Hyperion Microscope using KBr pellet technique. The reason for using KBr in FTIR analysis was its wide spectral range.

## 2.3  Morphological Characterizations of Synthesized HAp

SEM and TEM micrographs revealed the morphology, particle size, and nature of particle distribution of the synthesized HAp.

### 2.3.1 Scanning Electron Microscopy (SEM)

Scanning electron microscopy provides images of surface by using focused beam of electron. SEM is used to evaluate grain size, particle size, and material homogeneity. Scanning electron microscope of make Quanta 200F and model FEI was used to examine the morphology of synthesized HAp. The instrument has a magnification range of minimum 12× to maximum 100,000×. Sample was coated by a thin conductive layer of gold using gold sputter machine. A sputtering machine of make Cressington-108 auto and Model No.: 7006-8 was used for gold coating of samples using a vacuum pump. The samples were analyzed at 20 kV with a magnification range of 500–30,000×.

### 2.3.2 Transmission Electron Microscopy (TEM)

TEM analysis accounts for a beam of electron to be directed on a specimen. The thickness of the specimen is basically less than 100 nm or in the form of grid. The interaction of the beam of electrons on the surface of the specimens generates images. TEM has comparatively higher resolution compared to other microscopy that enables it to capture finer details even at atomic levels.

The morphology and the microstructure of the HAp derived from Bovine bones were investigated using transmission electron microscope of Make JEM-100 CX II. The machine operates at 100 kV. The particle size was calculated from TEM images using ImageJ software. It is an open-source software used for multidimensional images. It can measure distances and angles, display, edit, analyze, and process images.

## 2.4 Thermal Characterization of the Synthesized HAp

The high thermal stability of the HAp derived from Bovine bone was confirmed from TG analysis.

### 2.4.1 Thermo Gravimetric Analysis (TGA)

The thermal stability of the HAp was examined through TG analysis. Basic principle of TGA is to measure the mass of substance as a function of time. The analysis is carried out to evaluate changes in mass, thermal stability, oxidation/reduction behavior, and decomposition [16]. TGA was conducted on NETZSCH STA 449 F3 Jupiter in the mixed environment of nitrogen and oxygen to investigate the weight loss due to heating. To ascertain the test, 20 mg HAp was heated from 30 to 900 °C to investigate the weight loss on heating. The rate of heating was maintained at 10 °C/min.

# 3  Results and Discussion

## 3.1  X-Ray Diffraction Analysis

XRD spectrum of the synthesized powder is shown in Fig. 3. The diffraction patterns reveal HAp was the only phase found in the calcined powder. The peaks were sharp which indicates the highly crystalline nature of HAp. The extracted HAp from bovine bones showed an average crystallite size of 30.39 nm which is smaller than the ones obtained by Khoo et al. [12]. Further, the synthesized HAp showed 80.4% crystallinity which corresponds to a high degree of crystal formation [11]. Since smaller crystallite size with high degree of crystallinity is much desirable in HAp for tissue engineering applications [17], therefore it can be concluded that acid pre-treatment of bovine bone results in better deproteinization which enhances the crystallographic properties of the extracted HAp. Similar observation was reported by Deb et al. [18], which reveals that acid pre-treatment of fish scales yields HAp with better properties.

## 3.2  Fourier-Transform Infrared Spectroscopy Analysis

FTIR spectrum of the synthesized powder is shown in Fig. 4. The spectra showed peaks at 1530 and 1650 cm$^{-1}$ which confirmed the presence of carbonate ions $(CO_3^{2-})$ [19]. The absorbance peak at 570 and 1021 cm$^{-1}$ corresponds to the $\upsilon4$ and $\upsilon3$ P–O stretching vibration of $PO_4$ group, respectively [20]. Similarly, the band at 3570 cm$^{-1}$ attributed to the characteristic stretching modes of hydroxyl groups (O–H) [21]. The presence of the absorbance peaks corresponding to phosphate,

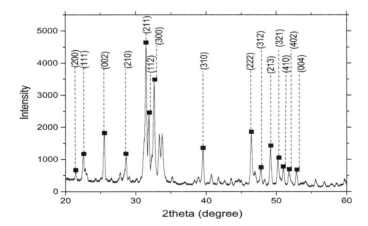

**Fig. 3**  XRD spectrum of bovine bone HAp

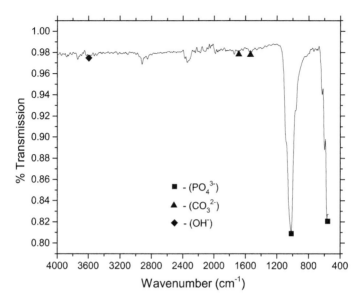

**Fig. 4** FTIR spectra of synthesized HAp

carbonate, and hydroxyl group confirmed the formation of hydroxyapatite in the calcined powder.

## 3.3 Scanning Electron Microscopy Analysis

SEM micrograph of synthesized HAp ceramic is represented in Fig. 5a. Formation of needle-like HAp flakes was observed from the SEM image on calcination at 1000 °C. Similar observation of HAp flake formation was also reported by Barakat

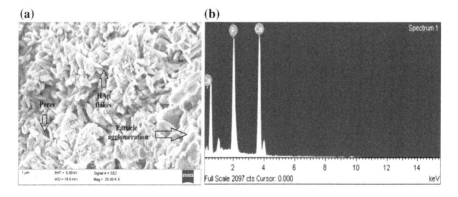

**Fig. 5 a** SEM micrograph and **b** EDX spectra of synthesized HAp

et al. [22]. Particle agglomeration with porous morphology was also observed for the synthesized HAp which is due to the high calcination temperature during HAp synthesis [23]. The formation of porous morphology and agglomerated HAp particle is best suited for biomedical applications. EDX spectra of the HAp shown in Fig. 5b further confirmed the existence of Ca and P in the calcined powder. The ratio of Ca/P observed from EDX analysis was found to be 1.71 which is fairly accurate with the stoichiometric ratio (1.67) of the pure HAp.

## 3.4   Transmission Electron Microscopy Analysis

TEM micrograph of the bovine-bone-derived HAp is depicted in Fig. 6 which showed rod-shaped HAp particles. The particle size of the synthesized HAp measured using ImageJ software was found to be approximately 68 nm. Selected area electron diffraction (SAED) pattern for the extracted HAp showed bright spots with concentric rings which indicate the presence of ultra-fine crystalline powder particles. This implied polycrystalline nature of HAp [24].

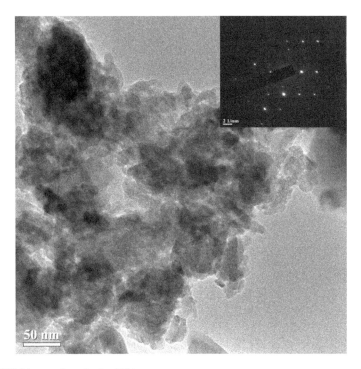

**Fig. 6** TEM image of synthesized HAp

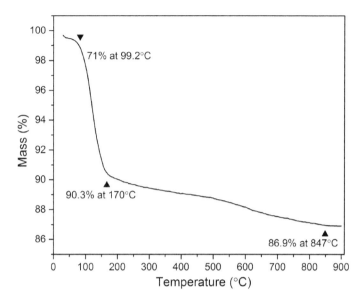

**Fig. 7** TGA curve for the synthesized HAp

## 3.5 TGA Results

TGA curve corresponding to weight loss of the HAp with respect to temperature is represented in Fig. 7. A minor degradation was observed up to 170 °C which is due to the presence of entrapped moisture. Minor degradation was further observed up to 847 °C that might correspond to the presence of any traces of organic matter. However, above 847 °C, there was almost no degradation of HAp which confirmed high thermal stability of the synthesized HAp [25]. It was also observed that there was only 13.1% degradation of HAp keeping residual mass of 86.9% that further confirmed high thermal stability of the synthesized HAp.

## 3.6 Quantification of Extracted HAp

The yield of the extracted HAp was calculated by measuring the dry weight of the raw bone considered before the process and the HAp extracted from it. Details of the weight measured during the formation of HAp from bovine bone are depicted in Table 1.

**Table 1** Percentage yield of the HAp synthesized from bovine bone

| Weight of raw dry bone (g) | Weight after cleaning with acetone (g) | Weight after deproteinization (g) | Weight after calcination (HAp obtained) (g) | % yield |
|---|---|---|---|---|
| 35 | 29.683 | 19.587 | 8.159 | 23.31 |

## 4  Conclusions

In the present study an attempt is made to synthesize nano-sized hydroxyapatite from bovine bone bio-waste by acid pre-treatment of bones followed by calcination at 1000 °C. The calcined powder obtained was characterized by FTIR and XRD analysis that confirms the successful formation of HAp. Intense and sharp peaks of XRD pattern imply crystalline nature of the synthesized HAp. Smaller crystallite size of 30.39 nm with 80.4% crystallinity shows enhanced crystallographic properties of the extracted HAp for biomedical applications. Presence of carbonate, phosphate, and hydroxyl ions corresponds to the formation of HAp in the calcined powder. Agglomerated HAp flakes with porous morphology are observed from SEM micrograph. TEM image reveals the formation of rod-shaped HAp particle with average size of 68 nm. SAED pattern indicates concentric rings with shiny spots confirming polycrystalline nature of HAp. High thermal stability with no degradation above 700 °C is shown in TGA curve for the synthesized HAp. Thus, acid pre-treated bovine bones yield HAp with enhanced properties that can have wide applications in the development of artificial bone material, bone void fillers for orthopedic, coating of orthopedic implants, drug delivery and in various other biomedical applications.

**Acknowledgements**  The authors acknowledge Material Science Laboratory of Mechanical Engineering Department, NIT Silchar, for performing TG analysis. The authors thank SAIF, IIT Madras, Chennai, for performing SEM and FTIR analysis. The authors are grateful to SAIF, Gauhati University and CIF, NIT Silchar for XRD analysis of samples. The authors thank Indovation Laboratory and TEQIP III, NIT Silchar, for providing fund for the characterizations of the samples.
**Ethical approval**  No Human/animal testing was performed during this study.

## References

1. Deb P, Deoghare AB, Borah A, Barua E, Lala SD (2018) Scaffold development using biomaterials: a review. Mater Today: Proc 5(5):12909–12919
2. Barua E, Deoghare AB, Deb P, Lala SD (2018) Naturally derived biomaterials for development of composite bone scaffold: a review. IOP Conf Ser: Mater Sci Eng 377(1):012013
3. Wang W, Yeung KWK (2017) Bioactive materials bone grafts and biomaterials substitutes for bone defect repair: a review. Bioact Mater 2(4):224–247
4. Bhui AS, Singh G, Sidhu SS, Bains PS (2018) Experimental investigation of optimal ED machining parameters for Ti-6Al-4 V biomaterial. FU Mech Eng 16(3):337–345
5. Lin K, Chang J (2015) Structure and properties of hydroxyapatite for biomedical applications. HAp Biomed Appl 4214(8):3–19

6.  Muhammad N, Gao Y, Iqbal F, Ahmad P, Ge R, Nishan U, Rahim A, Gonfa G, Ullah Z (2016) Extraction of biocompatible hydroxyapatite from fish scales using novel approach of ionic liquid pretreatment. Sep Purif Technol 161(7):129–135
7.  Mondal S, Mondal A, Mandal N, Mukhopadhyay SS, Dey A, Singh S (2014) Physico-chemical characterization and biological response of Labeo rohita-derived hydroxyapatite scaffold. Bioprocess Biosyst Eng 37(7):1233–1240
8.  Ho WF, Hsu HC, Hsu SK, Hung CW (2013) Calcium phosphate bioceramics synthesized from eggshell powders through a solid state reaction. Ceram Int 39(6):6467–6473
9.  Panda NN, Pramanik K, Sukla LB (2014) Extraction and characterization of biocompatible hydroxyapatite from fresh water fish scales for tissue engineering scaffold. Bioprocess Biosyst Eng 37(3):433–440
10. Rajesh R, Hariharasubramanian A, Ravichandran YD (2012) Chicken bone as a bioresource for the bioceramic (Hydroxyapatite). Phosphorous, Sulfur Silicon Relat Elem 187(8):914–925
11. Rana M, Akhtar N, Rahman S, Jamil HM, Asaduzzaman SM (2017) Extraction of hydroxyapatite from bovine and human cortical bone by thermal decomposition and effect of gamma radiation: a comparative study. Int J Comple Altern Med 8(3):00263
12. Khoo W, Nor FM, Ardhyananta H, Kurniawan D (2015) Preparation of natural hydroxyapatite from bovine femur bones using calcination at various temperatures. Proc Manuf 2:196–201
13. Gautam CR, Tamuk M, Manpoong CW, Gautam SS, Kumar S, Singh AK, Mishra VK (2016) Microwave synthesis of hydroxyapatite bioceramic and tribological studies of its composites with $SrCO_3$ and $ZrO_2$. J Mater Sci 51(10):4973–4983
14. Landi E, Tampieri A, Celotti G, Sprio S (2000) Densification behaviour and mechanisms of synthetic hydroxyapatites. J Euro Ceram Soc 20:2377–2387
15. Pang YX, Bao X (2003) Influence of temperature, ripening time and calcination on the morphology and crystallinity of hydroxyapatite nanoparticles. J Euro Ceram Soc 23(10):1697–1704
16. Bains PS, Payal HS, Sidhu SS (2017) Analysis of coefficient of thermal expansion and thermal conductivity of bi-modal SiC/A356 composites fabricated via powder metallurgy route. https://doi.org/10.1115/ht2017-5122
17. Boudemagh D, Venturini P, Fleutot S, Cleymand F (2018) Elaboration of hydroxyapatite nanoparticles and chitosan/hydroxyapatite composites : a present status. Polym Bull. https://doi.org/10.1007/s00289-018-2483-y
18. Deb P, Deoghare AB (2019) Effect of pretreatment processes on physicochemical properties of hydroxyapatite synthesized from *Puntius conchonius* fish scales. Bull Mater Sci 42(3):1–9
19. Zhang Y, Yokogawa Y (2008) Effect of drying conditions during synthesis on the properties of hydroxyapatite powders. J Mater Sci Mater Med 19(2):623–628
20. Destainville A, Champion E, Laborde E (2003) Synthesis, characterization and thermal behavior of apatitic tricalcium phosphate. Mater Chem Phys 80(1):269–277
21. Wei M, Evans JH, Bostrom T, Grondahl L (2003) Synthesis and characterization of hydroxyapatite, fuoride-substituted hydroxyapatite and fuorapatite. J Mater Sci Mater Med 14(4):311–320
22. Sheikh FA, Yong H (2009) Extraction of pure natural hydroxyapatite from the bovine bones bio waste by three different methods. J Mater Process Technol 209:3408–3415
23. Manalu JL, Soegijono B, Indrani DJ (2015) Characterization of hydroxyapatite derived from bovine bone characterization of hydroxyapatite derived from bovine. Asian J Appl Sci 3(4):758–765
24. Xu JL, Khor KA, Dong ZL, Gu YW, Cheang P (2004) Preparation and characterization of nano-sized hydroxyapatite powders produced in a radio frequency (rf) thermal plasma. Mater Sci Eng, A 374(1–2):101–108
25. Mondal S, Mahata S, Kundu S, Mondal B (2010) Processing of natural resourced hydroxyapatite ceramics from fish scale. Adv Appl Ceram 109(4):234–239

# Chapter 11
# Manufacturing and Evaluation of Corrosion Resistance of Nickel-Added Co–30Cr–4Mo Metal Alloy for Orthopaedic Biomaterials

Amit Aherwar

## 1 Introduction

Today, the major problem faced by a doctor and researcher in the field of orthopaedic materials is the selection of right biomaterials with correct proportions. Metallic materials such as stainless steel (SS), cobalt–chromium alloys (Co–Cr) and titanium and titanium alloys (Ti) are the effective orthopaedic materials used for implants. Table 1 lists the various materials including metallic, composites, ceramic and so on are used in orthopaedic application [1–3], and Table 2 lists the various possible material combinations of orthopaedic implants [4, 5]. These combinations of metallic biomaterials as listed in Table 2 have good corrosion resistance, good mechanical properties and biocompatibility, which make them a splendid choice for orthopaedic applications [6]. However, instead of these properties, there is some flaw in the metallic orthopaedic materials such as high elastic modulus, which causes stress shielding and corrosive nature. Corrosion deteriorates the implant materials in the form of metal ions and these ions liberated into the tissue resulting in adverse reactions [7]. 316 and 316L grades of stainless steel are the prime grades utilized to manufacture artificial bone as it is easy to cast into distinct shapes and sizes. Grade 316L has a healthier corrosion resistance as compared to 316 grades due to the attendance of less percentage of carbon content in the matrix alloy. Both 316 and 316L grades of stainless steel are easy to make fracture plates, screws, and hip nails. Due to ease of fabrication and desirable assortment of mechanical properties, corrosion behaviour, stainless steel becomes the predominant implant alloy [8]. Based on the superior results, ASTM has strongly recommended 316L grade as a foremost alloy for implant production [9, 10]. However, one more metallic material such as cobalt-based alloys are amid the most favourable orthopaedic biomaterials for making implants components such as

A. Aherwar (✉)
Mechanical Engineering Department, Madhav Institute of Technology and Science, Gwalior 474005, India
e-mail: amit.aherwar05@gmail.com; amit.aherwar05@mitsgwalior.in

© Springer Nature Singapore Pte Ltd. 2019
P. S. Bains et al. (eds.), *Biomaterials in Orthopaedics and Bone Regeneration*, Materials Horizons: From Nature to Nanomaterials, https://doi.org/10.1007/978-981-13-9977-0_11

**Table 1** Various materials used in orthopaedic application

| Materials | | Applications | References |
|---|---|---|---|
| Metals | SS-316L | Used in femoral head and stem components | [7, 8, 12] |
| | Co alloys | Used in femoral heads and stems, porous coatings, tibial and femoral components | |
| | Cast CoCrMo<br>Wrought Co–Ni–Cr–Mo<br>Wrought Co–Cr–W–Ni | | |
| Titanium-based materials | CP Ti | Used in porous coatings second phase in ceramic and PMMA composites | [10] |
| | $Ti_6Al_4V$ | Used in femoral heads and stems, porous coatings, tibial and femoral components | |
| | $Ti_5Al_{2.5}Fe$ | Used in femoral head and stem components | |
| | Ti–Al–Nb | Used in femoral head and stem components | |
| Ceramics | Bioinert | | [24–26] |
| | Carbon | Used in metal coatings on femoral stem components, second phase in composites and bone cement | |
| | Alumina | Used in femoral stems heads and acetabular cup components | |
| | Zirconia | Used in femoral stems and acetabular cup components | |
| Polymers | PMMA | Used in acetabular cups, tibial and patellar components | [27] |
| | UHMWPE/HDPE | Used in porous coatings on metallic and ceramic femoral stem components | |
| | Polysuffolene | Used in femoral stems, porous coatings on metallic femoral stem components | |
| | PTFE | Used in femoral stems, porous coatings on metallic femoral stem components | |

**Table 2** Materials' combination used in orthopaedic hip implants

| Material used for Femoral-Socket component | Results | References |
|---|---|---|
| CoCrMo–CoCrMo | Aseptic loosening rate is high, restricted use, minimum wear rate | [19, 21–23] |
| CoCrMo-UHMWPE | Widely in use, minimum wear loss | [21–23] |
| Alumina/zirconia-UHMWPE | Minimum wear rate | [10, 24–26] |
| Alumina–Alumina | Low wear rate, problem of joints pain | [10, 24] |
| $Ti_6Al_4V$-UHMWPE | Maximum wear | [27] |
| Surface-coated $Ti_6Al_4V$-UHMWPE | Better wear resistance to abrasion, attained skinny layer | [27] |

hip and knee joints, owing to their excellent mechanical strength and corrosion resistance [11]. Co–Cr alloy is commonly utilized as implants and fixations due to much attuned with the human body. In contrast with further biomaterials, Co–Cr alloys have a good biocompatibility property than SS (Both 316 and 316L) alloys but lesser than titanium alloys. Basically, Co-based alloys can be cast, wrought, or forged [12]. Accumulation of nickel in the modified $Co_{30}Cr_4Mo$ alloy enhances the corrosion properties due to slow rate oxidation and improved the mechanical properties [13, 14]. The density of pure nickel is 8.89 $g/cm^3$; melting temperature is 1468 °C and an elastic modulus of 209 GPa. Nickel specifically stabilizes the FCC structure and gives strength of the modified alloy. In fact, Nickel also improves the workability and castability of the alloy [15]. This element is considered corrosion-resistant due to its slow rate of oxidation in air at room temperature. Nevertheless, when metal alloys are used as implants and subjected to the body fluid, corrosion is inevitable. Corrosion is a continual process taking place on the surface of metals releasing ions into the surrounding media [16–20]. These ions may be biologically active and potentially carcinogenic [21, 22]. Therefore, all-inclusive knowledge of the effect of these ions on the surrounding tissues is requisite and attempts should be made to prevent the associated adverse effects. For example, the oxidation of Co–Cr–Mo routinely produces Co and Cr cations found in serum of patients with prosthesis, and molybdenum forms ionic species soluble in water; which are toxic [23, 24]. Depending on the nature and concentration of such chemical species, numerous inimical reactions may take place including allergy, infections, metallosis, skin toxicity, and many more [25, 26]. Therefore, Co–Cr–Mo alloy still lacks enough corrosion resistance to perform successfully in long-term use in the human body. This is of critical importance for implants with metal-on-metal (MoM) components such as MoM hip prostheses [27].

The authors in their earlier work [9] manufactured Co–30Cr alloys with modified molybdenum (Mo) content as potential biomaterials for hip implants and optimize it. Furthermore, analysis of variance (ANOVA) was also implemented to investigate the effect of molybdenum on mechanical and wear properties. From their results, it was concluded that the 4 wt% molybdenum alloying element provides minimal material loss. In the continuation of previous study, therefore, the present study manufactures the orthopaedic material consisting of $Co_{30}Cr_4Mo$ alloy as a base matrix and nickel as a filler to improve the properties of designed biomaterials and assess their properties with the focus on corrosion behaviour which has still remained as one of the most challenging clinical problems.

## 2  Test Materials and Methods

### 2.1  Test Materials

Mittal Industries, India, supplied the grades of raw materials such as Cobalt (Co), Chromium (Cr), Molybdenum (Mo) and Nickel (Ni) to make test specimens with distinct wt% Ni, in which cobalt was in ingot form and the rest were in powders form with size of below 44 µm. The micrographs of all the alloying elements are shown in Fig. 1. Furthermore, the element content with weight percentage of the manufactured alloys is listed in Table 3.

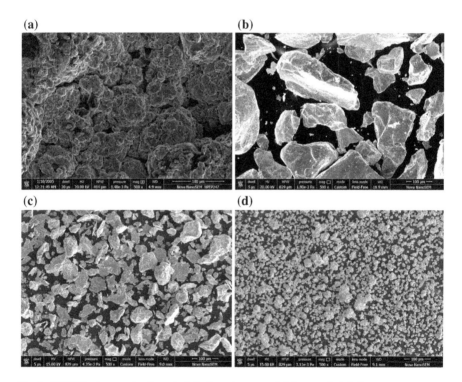

**Fig. 1** FESEM micrograph of **a** cobalt, **b** chromium, **c** molybdenum, **d** nickel

**Table 3** Element content and designation of specimens (in weight percentage)

| Sample designation | Elements | | | |
|---|---|---|---|---|
| | Co | Cr (%) | Mo (%) | Ni (%) |
| N0 | Bal | 30 | 4 | 0 |
| N1 | Bal | 30 | 4 | 1 |
| N2 | Bal | 30 | 4 | 2 |
| N3 | Bal | 30 | 4 | 3 |
| N4 | Bal | 30 | 4 | 4 |

## 2.2  Manufacturing of Orthopaedic Material

The schematic view of vacuum-based casting set-up used for manufacturing orthopaedic materials is shown in Fig. 2. In the presented work, five plates (100 mm × 65 mm × 10 mm) with x wt% of Ni ($x = 0, 1, 2, 3$ and 4) added Co–30Cr–4Mo alloy were produced using an induction furnace according to the composition. In this apparatus, there are two separate sections available in the set-up: (1) melting section and (2) casting section. A chiller unit is also attached in the set-up. Here, both the sections are under vacuum background. A motor was coupled with the bottom section and set to be around 200 rpm for proper mixing of all the metals present in the matrix. For manufacturing, all the proposed materials with respective weight percentages were melted above 1800 °C for 12 min and then dropped downwards into the graphite mould (100 mm × 65 mm × 10 mm) with the help of plunger under vacuum conditions. After casting, the mould was removed and then cut as per the sample size.

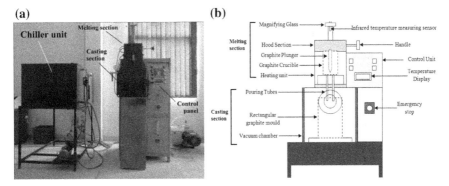

**Fig. 2** Experimental set-up of casting machine, **a** image of the casting set-up, **b** schematic diagram of the casting machine used for making the specimens

## 2.3  Material Characterization

The morphology of materials was characterized using FESEM and EDS of FEI Nova Nano SEM 450. The phase and crystal structure of the manufactured orthopaedic materials were studied using XRD and for the same PANalytical X'Pert PRO X-Ray diffractometer was utilized. The fabricated samples were polished by using Buehler MetaServ 250 polisher/grinder. After polishing, the specimens were etched for 12 s.

## 2.4  Mechanical Studies

Micro-hardness evaluation was carried out by USL micro-hardness tester. A diamond indenter was forced into the manufactured specimen under a load of 0.5 N for 10 s. The density of the manufactured alloys was recorded by the Archimedian principle. For measurement of physical and mechanical properties, three specimens were tested and the mean value was recorded for more precise results.

## 2.5  Electrochemical Test

The electrochemical behaviour and corrosion resistance of the manufactured alloys were investigated in the most unfavourable environment with aggressive pH (NaCl solution). It is unfeasible to accomplish corrosion tests on these orthopaedic biomaterials by the weight loss methodology since their rate of corrosion is acutely small and it takes a lengthy to obtain results. Thus, the foremost widespread method in corrosion studies, which is the recording of anodic polarization curves by the employment of the Tafel method, was used in this study. Electrochemical tests were conducted in a corrosion cell (see Fig. 3) using three electrodes; one was working electrode whose degradation property has to be tested, the second electrode was graphite as the auxiliary/counter electrodes and the third electrode used was saturated calomel electrode (SCE) which was used to sense the reaction happening within the corrosion cell and to input the values to the software where necessary data was shown in the form of graphs. GAMRY potentiostat VFP600 instrument was employed for performing electrochemical tests by accelerating corrosion with appropriate current and voltages. This instrument was allied to a PC and accessed with the assistance of Gamry Framework software. A potentiostat with corrosion software (Echem Analyst) was utilized for data analysis. The corrosion current density ($i_{corr}$) and other corrosion parameters were calculated from the polarization curves by Tafel extrapolation. Saline solution was prepared using 8 g/L NaCl, with pH 7.4 at the body temperature (37 °C). The capacity of the cell was 100 ml. The anodic polarization curve for the Tafel analysis was measured from $-150$ to $+150$ mV v/s $E_{corr}$ with a scanning rate of 1 mV/s after immersing in electrolyte about 30 min. The potentiodynamic curves

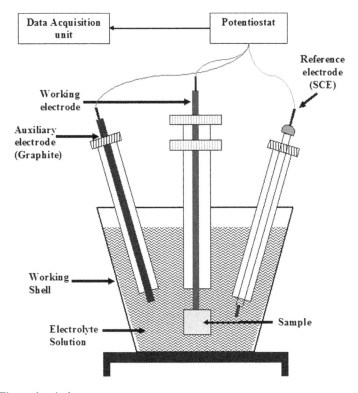

**Fig. 3** Electrochemical test set-up

were recorded at a constant voltage scan rate of 10 mV/s. The specimen area exposed
was 8 cm$^2$ with scan frequencies ranging from −1 to 1 V.

In order to compute the $i_{corr}$, usually, the Stern-Geary equation is used as shown
in Eq. (1). Since it was not possible to evaluate the coefficient value of Tafel's $b_a$ and
$b_c$, corrosion current was estimation using approximated Stern-Geary's Eq. (2).

$$i_{corr} = \frac{b_a \times b_c}{2.3(b_a + b_c)R_p} \tag{1}$$

$$i_{corr} = \frac{0.026}{R_p} \tag{2}$$

where $b_a$ is a slope coefficient of the anodic Tafel line; $b_c$ is a slope coefficient of the
cathodic Tafel line and $R_p$ is a polarisation resistance ($\Omega cm^2$).

Meanwhile, the surface micrographs of all samples after immersion in saline
(NaCl) solution at 37 °C with pH of 7.4 were provided for further evaluation.

# 3   Results and Discussion

## 3.1   Micro-structure and Phase Analysis of Materials

The XRD patterns of all the alloys are presented in Fig. 4. The micro-structure of
nickel-free alloy showed a cobalt matrix with chromium and molybdenum regions.
The HCP structure of Co has been formed due to martensitic transformation [28,
29]. None of carbide particles are shown in the micro-structure, which corresponds
healthy with the XRD peaks observed in 0–4 wt% Ni. Moreover, the compounds
cobalt (Co), chromium (Cr), molybdenum (Mo) and nickel (Ni) can be seen clearly
confirming its presence (Fig. 4) in the matrix. The lattice parameters which were
detected are provided in Table 4. The peaks recorded in this study are alike from the
previous studies [30–32]. Scanning electron micrographs and their corresponding
EDX results are shown in Fig. 5.

## 3.2   Mechanical Studies of Manufactured Material

The test results of the physical and mechanical strength of the manufactured
alloys with different nickel concentration (Co–30Cr–4Mo) are listed in Table 5.
The obtained results show that the density significantly increased with the

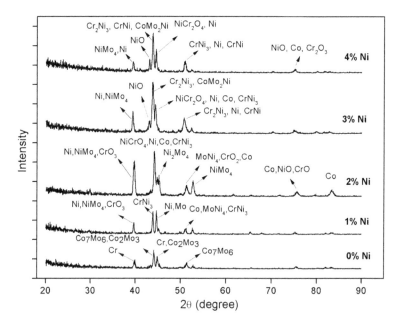

**Fig. 4**  XRD of the $Co_{30}Cr_4Mo$ alloy with 0–4 wt% of nickel content

**Table 4** Identified phases and the lattice parameters

| Phases | Crystal structure | Lattice parameters |
|---|---|---|
| Cobalt base $\alpha$ matrix | FCC (111) | $d = 2.04$ Å, $a = b = c = 3.545$ Å |
| Cr | BCC (110) | $d = 2.039$ Å, $a = b = c = 2.884$ Å |
| Mo | BCC (110) | $d = 2.225$ Å, $a = b = c = 3.147$ Å |
| $CrNi_3$ | Cubic (111) | $d = 2.0507$ Å, $a = b = c = 3.552$ Å |
| $Co_2Mo_3$ | Tetragonal (411) | $d = 2.0311$ Å, $a = b = 9.229$ Å and $c = 4.827$ Å |
| $Co_7Mo_6$ | Rhombohedral (116) | $d = 2.08$ Å |
| $Co_7Mo_6$ | Rhombohedral (027) | $d = 1.796$ Å $a = b = 4.762$ Å, and $c = 25.617$ Å |
| $MoNi_4$ | Tetragonal (002) | $d = 1.782°$A, $a = b = 5.724$ Å, $c = 3.564$ Å |
| $NiCr_2O_4$ | Cubic (220) | $d = 2.9407$ Å, $a = b = c = 8.318$ Å |
| $NiCrO_3$ | Rhombohedral (104) | $d = 2.648$ Å, $a = b = 4.925$, $c = 13.504$ Å |

incorporation of nickel concentration. The densities obtained from the test results are close to ASTM F75 [33].

## 3.3  Micro-hardness

The Vickers micro-hardness characteristics were measured at six distinct spots and the average value was taken. Table 5 represents the values of micro-hardness of the manufactured alloys. It was observed that the hardness of the nickel-added Co–30Cr–4Mo alloy increases linearly with the increase in nickel wt%. This may be quite evident and expected since hard nickel particles were mixed with base matrix and consequently contributed to effectively increase the hardness of the specimens (N1–N4). The utmost value was observed at 4 wt% nickel concentration, i.e. 740Hv. An improvement indicates the good bonding of nickel particles with modified Co–Cr–Mo alloys. A similar tendency was accounted by Savas and Alemdag [34] and Choudhary et al. [35] for the hardness of Al–40Zn–3Cu alloy and zinc–aluminium alloy with using nickel as a filler material, respectively. The authors investigated the influence of nickel (Ni) on the micro-structure and mechanical strength of the Zn–Al alloys.

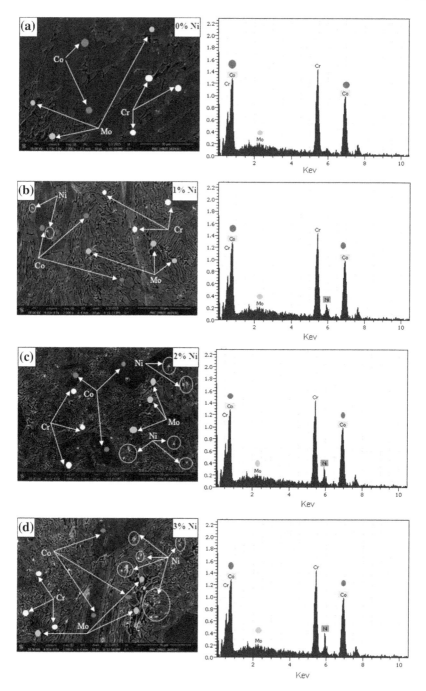

**Fig. 5** FESEM micro-structure and corresponding EDX results of fabricated material with different nickel content: **a** N0, **b** N1, **c** N2, **d** N3 and **e** N4. Cobalt (Co), chromium (Cr), molybdenum (Mo) and nickel (Ni) are identified. Ni is shown just by the small white spots in the micro-structure

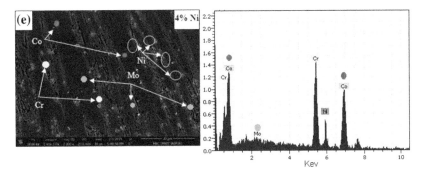

**Fig. 5** (continued)

**Table 5** Physico-mechanical properties of manufactured orthopaedic materials

| Sample | Density (gm/cc) | Standard deviation | Hardness (HV) | Standard deviation |
|--------|-----------------|--------------------|----------------|--------------------|
| N0 | 8.7 | 0.01866 | 762 | 0.07891 |
| N1 | 7.24 | 0.02175 | 590 | 0.48274 |
| N2 | 7.8 | 0.01784 | 640 | 0.47539 |
| N3 | 8.14 | 0.02156 | 690 | 0.18477 |
| N4 | 8.58 | 0.01888 | 738 | 0.29489 |

## 3.4 Polarization Curves and Corrosion Behaviour of Manufactured Material

This analysis revealed that the accumulation of nickel has strong impact on their corrosion behaviour. Figure 6a represents the open-circuit potential (OCP or $E_{OC}$) for the Co–30Cr–4Mo alloys with the addition of different wt% of nickel worked as stationary electrodes when they were submerged in the NaCl solution at 37 °C and pH of 7.4. In this approach, the OCP of a metal varies as a function of time but stabilizes at a fixed value after a lengthy spell of immersion. If open-circuit potential is higher, then the material is characterized as better corrosion-resistant. Based on that and obtained results in Fig. 6a, after 1 h exposure, an approximately constant value of −0.29 V was attained when adding 4 wt% of nickel content in the alloying composition which was higher (see Fig. 6a) compared with those obtained for N1, N2 and N3.

Figure 6b and Table 6 show the anodic polarization curves of the fabricated Co–30Cr–4Mo alloy with distinct wt% of Ni and their corresponding corrosion parameters, respectively. As it can be seen in Table 6, the $E_{corr}$ for 4 wt% of nickel-added alloy under NaCl solution was much more negative than those of 0–3 wt% Ni. Furthermore, the corrosion current of 4 wt% of nickel-added alloys was also much

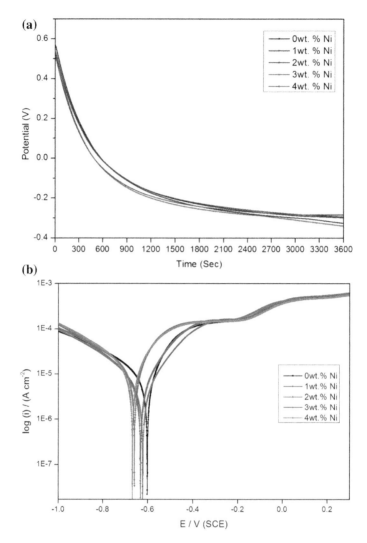

**Fig. 6** **a** Open-circuit potential (OCP) curves and **b** Tafel polarization curves for different wt% of nickel

**Table 6** Corrosion parameters for fabricated alloys at 37 °C with Tafel polarization method

| Sample | $I_{corr}$ ($\mu$A/cm$^2$) | $E_{corr}$ (mV) | Corrosion rate (mm/year) |
|---|---|---|---|
| N0 (Co$_{30}$Cr$_4$Mo + 0 wt% Ni) | 300 | −689 | 0.499 |
| N1 (Co$_{30}$Cr$_4$Mo + 1 wt% Ni) | 184 | −677 | 0.458 |
| N2 (Co$_{30}$Cr$_4$Mo + 2 wt% Ni) | 170 | −658 | 0.432 |
| N3 (Co$_{30}$Cr$_4$Mo + 3 wt% Ni) | 164 | −641 | 0.239 |
| N4 (Co$_{30}$Cr$_4$Mo + 4 wt% Ni) | 124 | −626 | 0.221 |

larger as compared to among the samples. The corrosion current ($I_{corr}$) is directly proportional to the corrosion rate, thus minimal the value of corrosion current ($I_{corr}$) the better protection against corrosion. Therefore, the alloy N0 was identified as lower corrosion-resistant material. Similarly, the more positive or anodic the corrosion current, the higher the protection against corrosion. Hence, the alloy N4 was indicated to have higher corrosion resistance material. By definition, the corrosion of implants implies that a certain amount of ions is released from the implanted metal into the body. The liberated ions may be settled down in certain parts of the body causing biological reactions [36]. When the corrosion potentials and currents of prepared materials with different wt% nickel are compared, it can be seen that Co–30Cr–4Mo alloy with addition of 4 wt% of nickel in the material composition has a more positive corrosion potential and lower corrosion current as compared to the other weight percentage of nickel content alloys. Hence, it can be concluded that Co–30Cr–4Mo with the addition of 4 wt% of nickel is more stable against corrosion than others in biological media.

Figure 7 shows the surface micrographs for nickel-free and nickel-added Co–30Cr–4Mo alloys after immersion in saline (NaCl) solution at 37 °C with 7.4 pH. The morphology of specimen surface in Fig. 7a reveals that in the absence of nickel content, the surface was highly corroded (shown in black spot) with areas of localized corrosion. However, in the presence of the nickel content (see Fig. 7b) in the matrix material, the rate of corrosion was suppressed; this can be observed from the decrease of corroded areas (black spots). Again, with further increase in nickel content in the matrix material, i.e. 2 wt% Ni, the corroded area was further reduced; the small black spots in Fig. 7c. This is an expected response because the corrosion current, i.e. $I_{corr}$ was low, i.e. ~170 $\mu$A/cm$^2$ (see Table 6), as compared to the other compositions. The lower is the corrosion current, the better its resistance to corrosion. By adding 3 wt% nickel to the alloy, the black spots (corroded surface) were further decreased (see Fig. 7d). However, addition of nickel in the alloy content, particularly 4 wt%, causes lower and smaller corroded areas (Fig. 7e). This is an expected response because the corrosion current, i.e. $I_{corr}$ was low, i.e. ~124 $\mu$A/cm$^2$ (see Table 6) and less amount of ions was released and caused more resistance against corrosion.

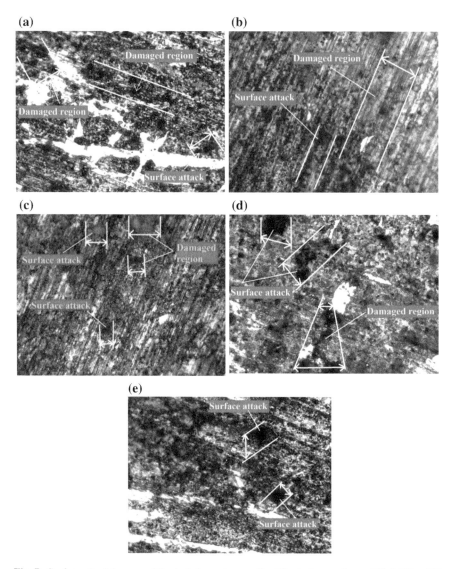

**Fig. 7** Surface attack images of the tested specimens after 90 min immersion: **a** N0, **b** N1, **c** N2, **d** N3 and **e** N4 alloy

# 4 Conclusions

This is the first study on evaluation of the influence of nickel on mechanical strength and corrosion resistance of the recently developed Co–30Cr–4Mo alloy. It was found that the mechanical properties of the orthopaedic material enhanced after adding certain amount of nickel content in a beneficial way. From this study, it can be concluded that electrochemical data analysis and surface micrographs of manufactured orthopaedics materials containing with 4 wt% nickel exhibit superior corrosion resistance than others, hence recommended for hip implant materials. However, further research could assess the developed materials in other biological solutions such as serum and joint fluid.

# References

1. Bhat SV (2005) Biomaterials, 2nd edn. Alpha Science International Ltd., Harrow
2. Alvarado J, Maldonado R, Marxuach J, Otero R (2003) Biomechanics of hip and knee prostheses. Applications of engineering mechanics in medicine. GED—University of Puerto Rico Mayaguez, pp 6–22
3. Bains PS, Mahajan R, Sidhu SS, Kaur S (2019) Experimental investigation of abrasive assisted hybrid EDM of Ti-6Al-4V. J Micromanuf. https://doi.org/10.1177/2516598419833498
4. Long M, Rack HJ (1998) Titanium alloys in total joint replacement-a materials science perspective. Biomater 19:1621–1639
5. Holzwarth U, Cotogno G (2012) Total hip arthroplasty. JRC Sci Policy rep. https://doi.org/10.2788/31286
6. Aherwar A, Singh A, Patnaik A (2016) Cobalt based alloy: a better choice biomaterial for hip implants. Trends Biomater Artif Organs 30(1):50–55
7. Cales B (2000) Zirconia as a sliding material-Histologic, laboratory, and clinical data. Clin Orthop Relat Res 379:94–112
8. Aherwar A, Singh A, Patnaik A (2016) Current and future biocompatibility aspects of biomaterials for hip prosthesis. J Bioeng 3(1):1–22
9. Aherwar A, Singh A, Patnaik A (2018) A study on mechanical behavior and wear performance of a metal-metal Co-30Cr biomedical alloy with different molybdenum addition and optimized using Taguchi experimental design. J Braz Soc Mech Sci Eng 40:213
10. Mirhosseini N, Crouse PL, Schmidth MJJ, Garrod D (2007) Laser surface micro-texturing of Ti–6Al–4V substrates for improved cell integration. Appl Surf Scie 253(19):7738–7743
11. Lewandowska-Szumiel M, Komender J, Chlopek J (1999) Interaction between carbon composites and bone after intrabone implantation. J Biomed Mater Res 48:289–296
12. Chang FK, Perez JL, Davidson JA (1990) Stiffness and strength tailoring of a hip prosthesis made of advanced composite materials. J Biomed Mater Res 24:873–899
13. Kauser F (2007) Corrosion of CoCrMo alloys for biomedical applications. University Birmingham, Metall Mater Sch Eng, pp 4–285
14. McMinn D, Daniel J (2006) History and modern concepts in surface replacement. Proc IMechE Part H: J Eng Med 220:239–251
15. Nomura N, Abe M, Kawamura A, Fujinuma S, Chiba A, Masahashi N, Hanada S (2006) Fabrication and mechanical properties of porous Co–Cr–Mo alloy compacts without Ni addition. Mater Trans 47(2):283–286
16. Bhui AS, Singh G, Sidhu SS, Bains PS (2018) Experimental investigation of optimal ED machining parameters for Ti-6Al-4V biomaterial. FU Mech Eng 16(3):337–345

17. Krasicka-Cydzik E, Oksiuta Z, Dabrowski J (2005) Corrosion testing of sintered samples made of the Co-Cr-Mo alloy for surgical applications. J Mater Sci Mater Med 16(3):197–202
18. Afolaranmi GA, Akbar M, Brewer J, Grant MH (2012) Distribution of metal released from cobalt–chromium alloy orthopaedic wear particles implanted into air pouches in mice. J Biomed Mater Res A 100(6):1529–1538
19. Moghaddam NS, Andani MT, Amerinatanzi A, Haberland C, Huff S, Miller M, Elahinia M, Dean D (2016) Metals for bone implants: safety, design, and efficacy. Biomanuf Rev 1:1
20. Bains PS, Grewal JS, Sidhu SS, Kaur S (2017) Wear between ring and traveler: a pin-on-disc mapping of various detonation gun sprayed coatings. Mater Today Proc 4(2):369–378
21. Rodrigues WC, Broilo LR, Schaeffer L, Knornschild G, Romel F, Espinoza M (2011) Powder metallurgical processing of Co-28% Cr-6% Mo for dental implants: Physical, mechanical and electrochemical properties. Powder Technol 206(3):233–238
22. Bahraminasab M, Hassan MR, Sahari BB (2010) Metallic biomaterials of knee and hip- a review. Trends Biomater Artif Organs 24(1):69–82
23. Hedberg Y, Wallinder IO (2014) Metal release and speciation of released chromium from a biomedical CoCrMo alloy into simulated physiologically relevant solutions. J Biomed Mater Res-Part B Appl Biomater 102(4):693–699
24. Panigrahi P, Liao Y, Mathew MT, Fischer A, Wimmer MA, Jacobs JJ, Marks LD (2014) Intergranular pitting corrosion of CoCrMo biomedical implant alloy. J Biomed Mater Res-Part B Appl Biomater 102(4):850–859
25. Kose N (2016) Biological response to orthopedic implants and biomaterials. In: Korkusuz FE (ed) Musculoskeletal res basic Sci. Springer, pp 3–14
26. Thakur RR, Ast MP, McGraw M, Bostrom MP, Rodriguez JA, Parks ML (2013) Severe persistent synovitis after cobalt–chromium total knee arthroplasty requiring revision. Orthop 36(4):e520–e524
27. Cadossi M, Mazzotti A, Baldini N, Gianini S, Savarino L (2016) New couplings, old problems: is there a role for ceramic-on-metal hip arthroplasty? J Biomed Mater Res Part B Appl Biomater 104(1):204–209
28. Ratner BD, Bankman I (2009) Biomedical engineering desk reference. Acad Press, Oxford
29. Montero-Ocampo C, Juarez R, Salinas-Rodriguez A (2007) Effect of FCC-HCP phase transformation produced by isothermal aging on the corrosion resistance of a Co-27Cr-5Mo-0.05C alloy. Metall Mater Trans A 33:2229–2235
30. Patel B, Inam F, Reece M, Edirisinghe M, Bonfield W, Huang J, Angadji A (2010) A novel route for processing cobalt–chromium–molybdenum orthopaedic alloys. J R Soc Interface 7:1641–1645
31. Rosenthal R, Cardoso BR, Bott IS, Paranhos RPR, Carvalho EA (2010) Phase characterization in as-cast F-75 Co–Cr–Mo–C alloy. J Mater Sci 45:4021–4028
32. ASTM F1537 (2000) Standard specification for wrought cobalt-28 chromium-6 molybdenum alloys for surgical implants. West Conshohocken, PA: ASTM Int
33. ASTM F75 (2014) Standard specification for cobalt-28 chromium-6 molybdenum alloy castings and casting alloy for surgical implants (UNS R30075). ASTM Int, Annual Book of Standards, West Conshohocken
34. Savas T, Alemdag Y (2010) Effect of nickel additions on the mechanical and sliding wear properties of Al–40Zn–3Cu alloy. Wear 268:565–570
35. Choudhury P, Das S, Datta BK (2002) Effect of Ni on the wear behaviour of a zinc-aluminium alloy. J Mater Sci 37:2103–2107
36. Basavarajappa S, Arun KV, Davim JP (2009) Effect of filler materials on dry sliding wear behavior of polymer matrix composites. J Min Mat Charact Eng 8(5):379–391

# Chapter 12
# Engineering of Bone: Uncovering Strategies of Static and Dynamic Environments

Jaya Thilakan, Ruchi Mishra, Sudhir K. Goel and Neha Arya

## List of Abbreviation

| | |
|---|---|
| ECM | Extracellular matrix |
| 3-D | Three dimension |
| HAp | Hydroxyapatite |
| Ti | Titanium |
| PLA | Polylactic acid |
| PGA | Polyglycolic acid |
| PLGA | Poly(lactic-*co*-glycolic acid) |
| PCL | Polycaprolactone |
| PLCL | Poly(lactide-co-$\varepsilon$-caprolactone |
| hADSCs | Human adipose-derived stem cells |
| BMSCs | Bone-marrow-derived MSCs |
| BSP | Bone sialoprotein |
| TCP | Tricalcium phosphate |
| ALP | Alkaline phosphatase |
| CPC | Calcium phosphate cement |
| hUCMSCs | Human umblical cord-derived mesenchymal stem cells |
| PVA | Polyvinyl alcohol |
| PEO | Polyethylene oxide |
| PAA | Polyacrylic acid |
| pDNA-NELL1 | Nel-like Type I molecular-1 DNA |

J. Thilakan · S. K. Goel · N. Arya (✉)
Department of Biochemistry, All India Institute of Medical Sciences Bhopal,
Saket Nagar, Bhopal 462020, Madhya Pradesh, India
e-mail: neha.biochemistry@aiimsbhopal.edu.in; neha.arya@gmail.com

R. Mishra
Department of Biotechnology, National Institute of Technology Raipur,
G.E. Road, Raipur 492010, Chhattisgarh, India

© Springer Nature Singapore Pte Ltd. 2019

175

P. S. Bains et al. (eds.), *Biomaterials in Orthopaedics and Bone Regeneration*,
Materials Horizons: From Nature to Nanomaterials,
https://doi.org/10.1007/978-981-13-9977-0_12

| hAFSCs | Human amniotic fluid-derived stem cells |
| OX2 | Osterix |
| RUNX2 | Runt-related transcription factor 2 |
| PEG | Polyethylene glycol |
| RGD | Arg-Gly-Asp |
| hESCd-MSC | Human embryonic stem cell-derived mesenchymal stem cells |
| PEGDA | Polyethylene glycol diacrylate |
| MSCs | Mesenchymal stem cells |
| HOB | Human osteoblast cells |
| ELR | Elastin-like recombinamer |
| hESCs | Human embryonic stem cells |
| iPSCs | Induced pluripotent stem cells |
| BMP-2 | Bone morphogenetic protein 2 |
| BMP-7 | Bone morphogenetic protein 7 |
| TPS | Tubular perfusion system |
| IL-1 | Interleukin-1 |
| IL-6 | Interleukin-6 |
| TNF-$\alpha$ | Tumour necrosis factor alpha |
| FGF-2 | Fibroblast growth factor 2 |
| M-CSF | Macrophage colony-stimulating factor |
| PDGF | Platelet-derived growth factor |
| BMPs | Bone morphogenetic proteins |
| VEGF | Vascular endothelial growth factor |
| TGF-$\beta$ | Transforming growth factor *beta* |
| IGFs | Insulin-like growth factors |
| bFBF | Basic fibroblast growth factor |
| LbL | Layer by layer |
| MMP | Matrix metalloproteinase |
| PD-MCG | Polydopamine-coated multichannel biphasic calcium phosphate granule system |
| BCP | Biphasic calcium phosphate scaffolds |
| CFD | Computational fluid dynamics |
| RPM | Rotations per minute |
| RWV | Rotating wall vessel |
| EMF | Electromagnetic field |
| PEMF | Pulsed electromagnetic field |
| GMP | Good manufacturing practice |
| Micro-CT | Microcomputed tomography |
| CAD | Computer-aided design |

# 1   Bone Tissue Engineering: An Introduction

As a highly specialized and dynamic tissue, bone is characterized by its mineralized matrix, rigidity and hardness with certain degree of elasticity. Bone provides support and protection to internal organs and also aids in locomotion. It is also involved in haematopoiesis and is an important reserve of calcium and phosphorus [1]. Further, the extracellular matrix (ECM) of bone comprises (a) non-mineralized organic phase, predominated by collagen type I, and (b) a mineralized inorganic phase, constituted of carbonated apatite [2]. This remarkable nanocomposite architecture confers properties such as high compressive strength and fracture toughness to bone. Additionally, there are other non-collagenous protein components that contribute to bone-specific events such as mineralization, osteoblast differentiation and bone remodelling [3].

During embryogenesis, bone formation either happens through intramembranous ossification or through endochondral ossification [4]. Once formed, at the morphological level, bone exists as compact (cortical bone) or spongy bone (cancellous bone). Cortical bone comprises tightly packed collagen fibrils that provide mechanical strength to bone. Cancellous bone, on the other hand, is a loosely arranged matrix that contributes to the metabolic functions of the bone. Being a highly dynamic tissue, bone undergoes constant modelling and remodelling [5]. During bone modelling, bone formation or resorption occurs on a bone surface. On the contrary, bone remodelling refers to bone formation or resorption that occurs sequentially in an organized manner with the aim to maintain structural integrity of the skeleton. Two major players involved in these processes are osteoblasts (bone-forming cells) and osteoclasts (bone-resorbing cells).

When subjected to trauma such as traumatic injury, soft tissue damage, tumour resection, age-related diseases such as osteoporosis or complications such as diabetes, bone responds by its ability to regenerate. However, if the defect size is greater than critical size, bone falls short of its regenerative capacity [6–10]. As a result, there is intervention in form of conventional treatment regimes such as autologous and allogenic grafts. Although autologous grafts remain the gold standard due to non-immunogenicity and recapitulating all properties of the requisite bone graft, it is limited by availability, donor site injury and morbidity [11]. Allogenic grafts from cadavers might be an alternative; however, it is associated with infections and graft rejection [12]. These issues led the researchers to harp on alternate bone repair strategies such as bone tissue engineering.

Bone tissue engineering aims to develop functional tissues and substitute the lost bone. In this context, tissue engineering encompasses three crucial elements, namely three-dimensional (3-D) transient structures called scaffolds, cells and growth factors, which drive the generation of a successful graft. Apart from these three elements, bioreactors have been shown to recapitulate certain aspects of in vivo bone microenvironment such as shear stress and mechanical stimulation [13–15]. This chapter will discuss the recent advances in the role of scaffolds, cells, growth factors and bioreactor-based strategies that have been studied towards the development of a successful graft.

## 2  Biomaterial-Based Scaffolds in Bone Tissue Engineering

Typically, scaffolds are porous, biodegradable and biocompatible materials with appropriate mechanical properties that facilitate adhesion, proliferation, differentiation and regeneration of damaged tissue [16]. For bone tissue engineering specifically, an ideal scaffold should be osteoinductive (able to induce osteogenesis by recruiting pre-osteoblasts/progenitor cells), osteoconductive (should support adhesion, proliferation and migration of osteoblasts throughout the construct) and osseointegrative (should be able to integrate into the surrounding bone). While a graft is osteoinductive, osteoconductive and osseointegrative, it should also recapitulate the mechanical properties of the surrounding anatomical site of implantation and maintain its integrity against the wear and tear caused during remodelling inside the host [17, 18]. Therefore, a biomaterial for the engineering of cortical or cancellous bone should be chosen based on their compressive strength that varies between 100–200 MPa and 2–20 MPa, respectively [19]. As an example, inspired by the excellent structural and mechanical properties of honey comb, Zhao and Liang [20] developed a 3-D-printed biomimetic comby scaffold using chitosan/hydroxyapatite (HAp) powder. Compressive strength ($1.62 \pm 0.22$ MPa) and Young's modulus ($110 \pm 22$ MPa) were found close to cancellous bone. Similarly, Chen et al. [21] fabricated biocompatible highly porous titanium (Ti) scaffold by powder metallurgy method with magnesium powder used as space holder; mechanical properties of the resulting scaffolds were close to human cortical bone.

In addition to the mechanical properties of the scaffold, bone regeneration is also influenced by the type of biomaterials, scaffold architecture and scaffold functionalization. Each of these will be discussed in the following section.

### 2.1  Choice of Biomaterials for Bone Tissue Engineering

According to the European Society of Biomaterials, a biomaterial is defined as "A material intended to interface with biological systems to evaluate, treat, augment or replace any tissue, organ or function of the body". Scaffolds used for bone tissue engineering are majorly categorized as **polymeric scaffolds, ceramic scaffolds, metallic scaffolds and their composites**. Figure 1 depicts examples of each of the aforementioned category [22–25].

Polymeric scaffolds are fabricated using either natural polymers (such as collagen, hyaluronic acid, agarose, chitosan and silk), synthetic polymers (such as polylactic acid (PLA), polyglycolic acid (PGA), poly (lactic-*co*-glycolic acid) (PLGA), polycaprolactone (PCL)) or their combination. Natural polymers hold superiority over synthetic polymer in terms of biocompatibility, biodegradability as well as biological information that supports cell attachment. However, they exhibit batch-to-batch variability, immunogenicity and risk of disease transfer as well as lack requisite mechanical strength. On the other hand, synthetic polymers are known for excellent

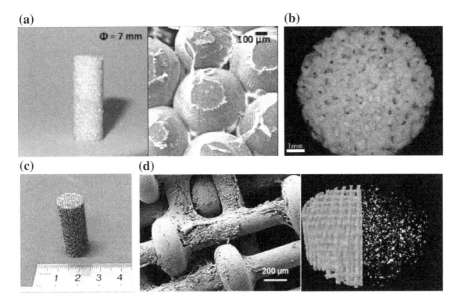

**Fig. 1** Biomaterials in bone tissue engineering; **a** digital photograph of silk fibroin scaffold (left) and SEM of silk fibroin scaffold showing silk fibroin microparticle arrangement in hexagonal fashion (right); **b** digital microscopic photograph of mesoporous bioactive glass; **c** digital photograph of porous Ti6Al4V scaffolds for bone tissue engineering. **d** SEM of human mesenchymal stem cell (hMSC)-seeded polylactic acid (PLA)/hydroxyapatite (HAp) composite scaffolds. Arrow depicts cell mineralization (left) and segmented microcomputed tomography image of PLA/HAp composite scaffolds depicting mineralized nodules (right)

mechanical properties and tailorability into various forms [26, 27] and however, they lack bioactive sites and hence are improvised by incorporating various bioactive molecules such as HAp or by blending with natural polymers [28]. As an example, in a study, silk fibroin (natural polymer) was blended with poly (lactide-co-$\varepsilon$-caprolactone) (PLCL) (synthetic polymer) in order to make the latter more conducive to cell–scaffold interaction [29]. It was found that blending of a natural polymer with a synthetic polymer supported differentiation of human adipose-derived stem cells (hADSCs) to osteogenic lineage. Furthermore, in vivo implantation of blended scaffold demonstrated enhanced bone volume, bone mineral density and new bone areas as compared to pure PLCL scaffold. In another study, human foetal osteoblasts were cultured on electrospun composite nanofibrous scaffolds based on HAp, chitosan and collagen in order to assess biomineralization. It was found that in comparison to HAp-chitosan scaffolds, collagen-doped composite scaffolds were highly biomimetic and osteoinductive [30]. To add to this, introduction of multiwalled carbon nanotubes to collagen–HAp composite scaffolds promoted the proliferation of bone-marrow-derived MSCs (BMSCs) as well as expression of osteogenic markers such as bone-sialoprotein (BSP) and osteocalcin as compared to collagen–HAp scaffolds [31].

Ceramic Scaffolds, specifically calcium phosphate scaffolds, such as tricalcium phosphate (TCP), Hap or their combination as biphasic systems are extensively studied in bone tissue engineering because they perfectly mimic the mineral composition of the bone and their biophysical properties make the scaffolds osteoconductive and osteoinductive [32–35]. Apart from the calcium phosphate ceramics, there are other bioactive ceramics such as calcium silicon-based ceramics termed akermanite (combination of calcium silicon and magnesium) and bioglass (combination of sodium oxide, calcium oxide, silicon dioxide and phosphorus pentoxide) that have also been explored for bone tissue engineering [36–38]. While calcium silicon-based ceramics are well known for their mechanical properties and controlled degradation rate, bioglass is known for its biodegradability and osteogenic potential [39]. However, brittleness of bioglass limits its use as a stand-alone scaffold and is therefore used as composites for bone regeneration [40].

Furthermore, scaffolds containing metals find application in bone regeneration due to properties such as high mechanical strength and fracture toughness [41, 42]. Commonly used metals are stainless steel 316L(ASTM F138), cobalt-based alloys (ASTM F75 and ASTM F799) and titanium-based alloys (Ti–6A1–4V, ASTM F67 and F136) [43]. However, metallic scaffolds exhibit poor biological recognition and release toxic metal ions due to corrosion or rusting, thereby resulting in allergic reactions and inflammation [44]. To overcome this, the surface of metallic scaffolds is usually modified or coated to improve biocompatibility; cell-recognizing ligands and growth factors have also been integrated within the proximity of the construct to enhance cell growth [45]. Surprisingly, a study demonstrated that Ti scaffolds have an intrinsic potential to promote osteogenesis. Briefly, BMSCs seeded on uncoated highly porous Ti scaffold and HAp-coated Ti scaffold were evaluated for their osteogenic potential. Results demonstrated that uncoated Ti scaffolds induced better bone formation and ingrowth when implanted in sheep stifle joints as compared to HAp-coated scaffolds, suggesting that Ti-based scaffolds are self-sustained to promote osteogenesis and have the potential to be used in healing large bone defects [46]. There are also reports of few other metals such as magnesium alloy W4 and copper-containing scaffolds in promoting osteogenesis [47, 48].

## 2.2  Role of Scaffold Architecture

Bone tissue engineering is influenced not only by the type of biomaterial but also by the design and geometry of scaffolds. Several studies have shown the influence of scaffold architecture in modulating cell behaviour and their differentiation [49]. This section will be discussing the influence of scaffold architectures, such as hydrogels, macroporous scaffolds and fibrous scaffolds, in bone tissue engineering (Fig. 2) [50–53].

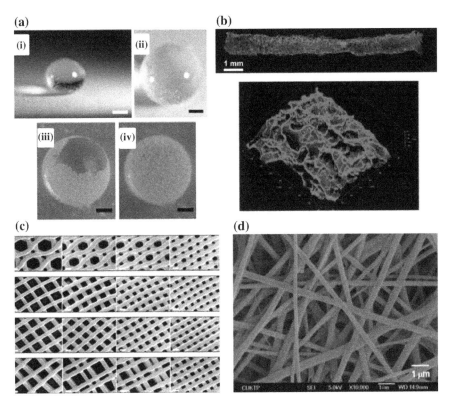

Fig. 2  Types of scaffolds in bone tissue engineering; **a** gelatin methacrylate drop on a hydrophobic surface coated with hydrophobic bioactive glass nanoparticles (i) and (ii). Droplet encapsulated with nanostructured film (iii), resulting in semitransparent liquid marble (iv). The polymer liquid core was then crosslinked with UV light resulting in bioactive hydrogel marble; **b** confocal micrographs of complete and surface of rhodamine-labelled scaffold based on pullulan, modified with a cholesterol moeity as obtained by freeze drying; **c** SEM micrographs of 3-D-printed bioactive glass ceramic (Sr-HT gahnite) scaffolds of different geometries; **d** SEM micrograph of poly (3-hydroxybutyrate-co-3-hydroxyvalerate) scaffolds generated by electrospinning

### 2.2.1  Hydrogels

Hydrogels are products of physical or chemical cross-linking between hydrophilic polymers (natural or synthetic) with the tendency to swell in biological fluids. Highly hydrated state of hydrogels makes them an ideal choice for cell encapsulation as well as differentiation [54, 55]. They also possess excellent capacity to entrap and release bioactive agents, thereby serving as promising candidates for bone regeneration. Various natural polymers such as collagen, gelatin, alginate, hyaluronic acid, agarose, chitosan and silk and synthetic polymers such as PLA, polyvinyl alcohol (PVA), polyethylene oxide (PEO), polyacrylic acid (PAA) and poly(propylene fumarate-co-ethylene glycol) have been used for hydrogel fabrication [56, 57]. Further, many hydrogels have been reported as injectable systems that can be introduced to the site of action by minimally invasive procedures and support in situ bone formation [58]. Amongst various injectable hydrogels reported, alginate is one of the well-studied biomaterials for bone tissue engineering [59]. In a study, Han et al. [60] prepared an injectable calcium silicate/sodium alginate composite hydrogel; in situ gelation was induced by calcium ions released from calcium silicate following the addition of D-gluconic acid $\delta$-lactone. The composite successfully induced HAp formation and promoted osteogenesis of rat BMSCs and angiogenesis of human umbilical vein endothelial cells. Other polymers such as chitosan, collagen and N-isopropylacrylamide have also been used as injectable systems for bone tissue engineering [61–63]. Additionally, composite hydrogels using HAp, nanosilica, bioglass and zinc have been utilized in bone tissue engineering to provide mechanical stability and promote in vivo calcification [64–70]. Furthermore, on the basis of type of cross-linking, injectable hydrogels can be thermoresponsive, photocrosslinked, chemically crosslinked or enzymatically crosslinked [71–73].

### 2.2.2 Macroporous Scaffolds

Macroporous scaffolds are highly interconnected, porous, 3-D structures that are based on natural or synthetic polymers. These can be generated using techniques such as particulate leaching, freeze drying, solvent casting, gas foaming, thermally induced phase separation and 3-D printing [74–76]. Various composite macroporous scaffolds have been reported for bone tissue engineering. In one such example, highly porous chitosan–silica composite scaffolds were fabricated by freeze drying; the scaffolds favoured osteoblast proliferation, with enhanced alkaline phosphatase (ALP) activity as well as mineral deposition in comparison to chitosan-only scaffolds [77]. Further, 3-D printing has also been used in bone tissue engineering [78–81] and allows for precise control over the architecture and geometry of the scaffold [82]. Commonly used biomaterials for 3-D printing of bone that can recapitulate the mechanical properties of bone and can promote vascularization include calcium phosphate composites, bioactive glass mixture, zirconium oxide, silica, graphene and strontium [83]. One such study utilized 3-D-printed porous scaffold based on bioactive glass and chitosan nanoparticles, loaded with an osteoinductive protein, Nel-like Type I molecular-1 DNA (pDNA-NELL1) and tested for efficacy of BMSCs to repair bone defects in rhesus monkey [81]. It was observed that enhanced alveolar bone regeneration was observed in 3-D-printed bioactive glass–chitosan nanoparticles loaded with pDNA-NELL1 and BMSCs as compared to other control groups.

Within macroporous scaffolds, pore size and interconnectivity are known to influence cell infiltration and bone regeneration [84]. In a study, bilayer HAp-based scaffold corresponding to cortical-cancellous organization in bone with pore size of 200 μm (as outer layer) and 450 μm (as inner layer) was compared to trabecular-like organization with a uniform pore size of 340 μm for their ability to support bone regeneration in a 10-mm segmental rabbit radius defect model. Result showed that uniform pore-sized scaffolds supported better functional bone, greater flexure strength as well as toughness when compared to scaffolds with bimodal pores [85].

In another interesting study, Gupte et al. [86] fabricated nanofibrous PLA scaffolds of controlled pore architecture using thermally induced phase separation and sugar porogen template method and evaluated BMSCs differentiation as a function of pore architecture. They found that smaller pore size (125–250 μm) favoured chondrogenesis while larger pore size (425–600 μm) supported mineralized bone tissue via the ingrowth of blood vessels within the porous architecture. This study suggested an important contribution of pore architecture in bone tissue engineering application.

Apart from pore architecture, few researchers have also explored the role of pore shape in modulating osteogenesis. As an example, Xu et al. [87] compared varying shapes (square, triangular and parallelogram) of porous ceramic scaffolds and found out that highest ALP activity was observed in scaffolds with parallelogram shape. While it is not very clear why parallelogram shape demonstrated highest ALP activity, nevertheless, it would be interesting to understand this at the mechanistic level.

### 2.2.3  Fibrous Scaffolds

Fibrous scaffolds have been widely explored for bone tissue engineering [88] since they mimic the fibrillar extracellular collagen networks within the bone. They can be fabricated via electrospinning, self-assembly or phase separation method. Electrospinning is one of the most versatile techniques since it can be used to generate scaffolds with controlled morphology (nanofibre diameter and orientation) as well as porosity [89]. A study reported the effect of electrospun fibre diameter and orientation on differentiation of BMSCs for bone repair. Results showed that BMSCs demonstrated more elongated and spindle-shaped morphology on aligned fibres compared to random fibres. Further, aligned, submicron/micron-sized fibres (906 ± 178 μm) showed higher expression of osteogenic markers such as Osterix (OSX) and Runt-related transcription factor 2 (RUNX2) as compared to other test groups (random fibres of diameter 1,183 ± 174 μm, aligned fibres of diameter 404 ± 107 μm and random fibres of diameter 449 ± 96 μm) [90]. Few others have also reported the effect of fibre orientation on enhanced osteogenesis [91–93].

## 2.3  Role of Scaffold Functionalization

Scaffold surface plays a crucial role in cell–scaffold interaction. In order to enhance a scaffold's bioactivity, its surface is modified to incorporate specific functions such as small functional groups, growth factors, small peptide sequences or complex bioactive molecules [94–96]. In a seminal work by Benoit et al. [97], phosphate-functionalized polyethylene glycol (PEG) surfaces led to osteogenic differentiation of hMSCs, thereby demonstrating that simple functional groups can be used to control complex cellular events. To add to this, Arora and Katti [98] also showed that phosphorylation and polysialylation of gelatin led to enhanced mineralized and osteogenic differentiation, respectively, on murine MSC line, C3H101/2. In another study, polyethyleneimine and citric acid-grafted 3-D-printed PLA scaffolds were subjected to simulated body fluid to generate calcium-deficient PLA-HAp. These scaffolds led to enhanced ALP activity and expression of various osteogenic markers as compared to PLA scaffolds demonstrating the role of surface modification on biological activity of PLA scaffolds [99]. Furthermore, conjugation of cell adhesion peptide Arg-Gly-Asp (RGD) to polymeric surfaces also modulates cell behaviour [100]. This was corroborated by a study performed by Chen et al. [101], wherein they investigated the osteogenic differentiation of human ESC-derived MSC (hESCd-MSC) on calcium phosphate cement–chitosan–RGD scaffolds. In vitro results showed significant attachment, proliferation and mineralization of hESCd-MSC when seeded on CPC–chitosan–RGD as compared to CPC–chitosan scaffolds.

Other moieties such as heparin have also been used to functionalize polymeric surfaces [102]. In one study, the effect of heparin functionalized chitosan scaffold on the activity of MC3T3-E1 pre-osteoblast cells was investigated. Result showed that scaffolds with covalently bounded heparin led to enhanced ALP and osteocalcin secretion in comparison to electrostatically bound heparin containing chitosan scaffold and heparin-free chitosan scaffold [103]. In another study, methacrylated polyethylene glycol diacrylate (PEGDA)/chondroitin sulphate-based hydrogels were subjected to chondroitin sulphate-mediated recruitment of ions like calcium and phosphate. Under in vitro conditions, human tonsil-derived MSCs seeded on biomineralized surfaces promoted expression of various osteogenic markers, and the hydrogel with 10% chondroitin sulphate demonstrated highest bone mineral density in critical-sized cranial defect model as compared to other conditions [104].

## 2.4   Cell Sources for Bone Tissue Engineering

Bone is a highly dynamic organ and comprises four active bone cell types—osteoblasts, osteoclasts, osteocytes and bone-lining cells [105]. Each of these cell types has a specific function contributing to the generation of a healthy bone, thereby suggesting that an ideal bone graft may require seeding of these cells on a scaffold. However, limited cell source is always a drawback for generation of such a graft [106]. Type of cells, their source and the protocol of cell seeding greatly influence the formation of any engineered tissue. Use of pre-differentiated osteoblasts from autologous source is the most obvious choice for bone tissue engineering, but their slow proliferation has led to use of alternate sources [107] which includes mesenchymal stem cells (MSCs) from bone marrow, adipose tissue and dental pulp, embryonic stem cells as well as genetically engineered osteogenic cells. A comparison between various cell sources has been depicted in Table 1. Clinical performance of these cells in bone tissue engineering would depend upon the ease of harvesting, in vitro expansion, in vivo osteogenesis, low/no immunogenicity as well as no transmission of pathogens. The following section will discuss the latest updates on various cell sources used in bone tissue engineering.

**Table 1** Advantages and disadvantages associated with cell sources for bone tissue engineering

| Cell type | Advantage | Disadvantage |
|---|---|---|
| Bone marrow-derived mesenchymal stem cells (BMSCs) | • High osteogenic potential<br>• Highly stable in culture media, even a small number of cells can yield high quantity<br>• Extensively studied | • Extracting procedure is highly invasive and painful<br>• Low yield at isolation and high risk of contamination<br>• Loss/decreased multipotency after extensive passages or when isolated from older age group |
| Adipose tissue-derived mesenchymal stem cells (ADSCs) | • Isolation is easy and less invasive<br>• Cell yield is higher than bone marrow aspirate and better genetic stability<br>• Low donor tissue morbidity | • High tendency of spontaneous differentiation into adipocytes |
| Dental pulp-derived mesenchymal stem cells (DPSCs) | • Highly proliferative with enhanced ability to differentiate into osteoblasts | • Low yield on isolation |
| Embryonic stem cells (ESCs) | • Highly pluripotent cells<br>• Unlimited self-renewal capacity | • Formation of teratomas<br>• Sufficient studies not available to enumerate stable and reproducible differentiation<br>• Immune incompatibility<br>• Ethical concerns |
| Induced pluripotent stem cells (iPSCs) | • Highly pluripotent cells with no ethical or immunological issues<br>• Can be generated through any cell source<br>• Highly patient-specific. | • Associated with tumorigenicity and spontaneous teratoma formation |
| Genetically engineered osteogenic cells | • Result in differentiation of engineered cells towards osteogenic lineage only and thus in enhanced bone regeneration | • Associated with immunogenicity, acute immunomodulatory effect and malignant transformation (due to uncontrolled insertional mutagenesis) |

### 2.4.1   Mesenchymal Stem Cells (MSCs)

MSCs show various properties such as self-renewal, multi-lineage differentiation and immunomodulation. As a result, they demonstrate potential as a promising tool in cell therapy and tissue engineering applications [108–110]. Stability of MSCs in the culture medium, ease of preparation and their potential to differentiate into osteoblasts present them as one of the most suitable candidates in the field of bone regeneration. Different sources for MSCs have been used in the past; this section will discuss major sources such as bone marrow-derived, adipose-derived and dental pulp-derived MSCs.

a. Bone-Marrow-Derived Mesenchymal Stem Cells (BMSCs)

BMSCs have gained enormous success in the recent past and have been extensively explored from the pre-clinical perspective. As an example, Nassif et al. [111] seeded BMSCs on chitosan scaffolds that were pre-treated with dexamethasone, an osteogenic inducer, and showed that BMSCs had osteoinductive properties when implanted on chitosan scaffolds under both in vitro and in vivo conditions as compared to empty scaffolds. Another study involved the encapsulation of BMSCs in matrigels which were then loaded on 3-D-printed porous titanium scaffold (Ti6Al4V), in order to provide appropriate mechanical strength. In vivo results showed that the scaffolds with BMSC-loaded matrigels showed better new bone formation in rats with full thickness critical mandibular defects in comparison to the rats treated with locally injected BMSCs scaffolds and pure matrigel-loaded scaffolds [112].

Though BMSCs have multilineage potential, their ability to differentiate into osteoblasts is the highest [113]. However, harvesting protocols for these cells is quite invasive and is also associated with a high risk of contamination, especially if the isolation is done from a patient having any other disease. Further, the number of cells isolated varies from patient to patient and declines with the age of the donor [114]. Additionally, strength of the population obtained is also governed by the isolation technique. Therefore, other cell types have also been explored for bone regeneration.

b. Adipose Tissue-Derived Mesenchymal Stem Cells (ADSCs)

ADSCs have appeared to be a successful alternative to BMSCs primarily due to the ease of extraction, abundance, rapid in vitro expansion and better genetic stability under in vitro conditions [115]. Many studies have explored ADSCs in bone tissue engineering applications. As an example, human ADSC-seeded collagen/HAp scaffolds were evaluated for ectopic bone formation following subcutaneous implantation in mice in comparison to cell-free scaffold. The results indicated augmented calcium deposition and vascularization in ADSC-seeded scaffolds as compared to cell-free scaffolds and demonstrated potential in cases of elderly or those with reduced regeneration capacity [116]. Another study compared the osteogenic capacity of human ADSCs and human osteoblast (HOB) cells on microchannel-patterned collagen–fibroin–elastin-like recombinamer (ELR) blend films. Although both ADSC and HOB-seeded constructs closely mimicked the ultrastructure of bone, ADSCs showed better osteogenic properties as compared to HOB when seeded on collagen–fibroin–ELR constructs, thereby demonstrating potential in bone tissue engineering [117].

ADSCs have appeared to be a popular source for bone tissue engineering applications. However, they demonstrate high tendency of spontaneous differentiation into adipocytes [118].

c. Dental Pulp-Derived Mesenchymal Stem Cells (DPSCs)

DPSCs have also been explored as an alternative to BMSCs due to enhanced proliferation rates. Moreover, DPSCs are a very homogenous population and have demonstrated enhanced ability to differentiate into osteoblasts, thereby demonstrating potential in bone tissue engineering [119]. In a study by El-Gendy et al. [120], potential of hDPSCs on 45S5 bioglass scaffolds to promote bone-like tissue formation under in vitro and in vivo conditions was tested. The authors found out hDPSCs promoted greater osteogenesis under basal as well as osteogenic conditions on 45S5 bioglass scaffolds as compared to cells grown on 2-D surfaces. Furthermore, peritoneal implantation of DPSC-seeded scaffolds demonstrated sporadic woven bone-like spicules as well as calcified tissue, showing potential in bone repair. Similar results have been reported in other systems; Petridis et al. [121] reported the healing response of cranial defects in rats when implanted with DPSC-seeded hyaluronic acid-based scaffolds. Results demonstrated superior bone regeneration in DPSC-seeded scaffold as compared to cell-free scaffolds. Even though DPSCs show potential in bone tissue engineering applications, full realization of clinical potential pertaining to DPSCs in bone tissue engineering requires the establishment of new strategies in this direction [122].

### 2.4.2 Pluripotent Stem Cells: Human Embryonic Stem Cells (hESCs) and Induced Pluripotent Stem Cells (iPSCs)

Pluripotent stem cells such as hESCs and iPSCs have also been explored as alternative cell sources for bone tissue engineering [123, 124]. In a study, Marolt et al. [125] showed that cultivation of hESCs on 3-D osteoconductive scaffolds in bioreactors (bioreactors are discussed later in this chapter) with interstitial flow of culture medium led to formation of compact, homogenous, stable bone-like tissue without differentiating into other lineages. In vivo implantation of engineered bone further resulted in maintenance and maturation of bone ECM without teratoma formation, a phenomenon constantly observed following implantation of undifferentiated hESCs. However, these cells exhibit ethical constraints since isolation of cells destroys the embryo. Further, on implantation, the cells invoke infiltration of inflammatory cells and subsequent rejection [126, 127].

Recently, iPSCs have been widely explored as a prospective cell source for bone tissue engineering [128]. In a study, human iPSCs cultured in macrochannelled PCL scaffolds demonstrated enhanced osteogenesis under in vivo conditions as compared to cell-free scaffolds [129]. Similar results were obtained when iPSC-derived osteoprogenitors were encapsulated in self-assembling peptide nanofibre hydrogels, followed by their implantation in a calvarial bone defect rat model [130]. While the iPSCs appear to be a promising source for bone regeneration, there are certain concerns especially tumorigenicity and spontaneous teratoma formation [131]. To address this concern, Xie et al. [132] used a new class of iPSCs called the iPSC-MSCs that act as a source of MSCs and at the same time are less tumourigenic compared to iPSCs [133]. The cells outgrowing from embryoid bodies that were in turn generated from iPSCs were designated as iPSC-derived MSCs. The derived cells were then seeded on HAp/collagen/chitosan-based biomimetic nanofibres and investigated for their bone regeneration in cranial bone defects in rats. It was revealed that iPSC-MSC/HAp/collagen/chitosan demonstrated nearly two-fold higher bone mineral density compared to other groups, and the system could be used as the new-stem cell–scaffold system in the field of bone tissue engineering.

### 2.4.3 Genetically Engineered Osteogenic Cells

Although MSCs have been profusely used in bone tissue engineering, their entry into distinct lineages comes at the cost of high concentration of exogenous growth factors [134]. In order to overcome this, researchers are genetically engineering MSCs with certain bone-specific genes in order to induce differentiation into osteogenic linage only, thereby demonstrating potential towards enhanced bone regeneration [135]. In a study by Huynh et al. [136], MSCs were genetically engineered for over-expression of RUNX2 with concomitant SMAD3 knockdown, and cell-seeded PCL scaffolds were tested for differential matrix deposition potential. Interestingly, genetically engineered MSCs demonstrated enhanced mineral deposition while the unmodified MSCs demonstrated enhanced glycosaminoglycan deposition, thereby showing application towards the regeneration of complex tissues. In another study, Kuttappan et al. [137] investigated the bone regeneration capabilities, more specifically in repair of segmental defects by using bone morphogenetic protein 2 (BMP-2)-engineered MSC-seeded composite scaffolds. The authors demonstrated that BMP-2-engineered MSCs showed better new bone formation in critical-sized rat femoral segmental defects as compared to non-transfected MSCs. In yet another study, Kargozar et al. [138] tried to accelerate the bone regeneration capacity of BMSCs by transfecting them with a plasmid containing bone morphogenetic protein-7 (BMP-7)-encoding gene. Both the modified and unmodified cells were seeded on bioactive glass/gelatin nanocomposite scaffolds, which were then evaluated for osteogenic potential in calvarial critical-sized defect in rats. In vivo results showed that higher rate of osteogenesis was observed in the group of animals implanted with modified BMSCs in comparison to cell-free scaffolds and group with unmodified BMSCs. Although genetically engineered osteogenic cells demonstrate enhanced osteogenic differentiation, they are associated with immunogenicity, acute immunomodulatory effects and malignant transformation (due to uncontrolled insertional mutagenesis).

Worldwide, series of clinical trials and studies are going on based on pluripotent stem cell-based therapies for the treatment of diseases such as mascular degeneration, and neurological disorders, haematological disorders and cardiovascular disorders; however, clinical application for bone regeneration is still underway [139, 140].

### 2.4.4   Co-culture Strategies for Bone Tissue Engineering

It is evident that bone is a complex construct having four active bone cell types, namely, osteoblasts, osteoclasts, osteocytes and bone-lining cells with well-defined functions, working in synchronization to achieve bone homoeostasis [105, 141]. Therefore, single-cell-type culture usually fails to mimic the bone microenvironment, thereby eliciting the need of co-culture strategies in bone tissue engineering. Moreover, an implant that is unable to generate vascularization inside the construct can lead to a necrotic graft further delaying the possibilities of clinical application.

In this regard, various co-culture models have been developed to understand the effect of the aforementioned cell types on bone regeneration [142]. As an example, Beskardes et al. [143] investigated the effect of perfusion co-culture based on osteoblast derived from MSCs and osteoclasts derived from THP-1, human acute monocytic leukaemia cell line on chitosan-HAp superporous scaffolds. On similar lines, Jeon et al. [144] recapitulated bone tissue remodelling by co-culturing osteoblast and osteoclasts derived from human iPSCs-MSCs and human iPSCs macrophages, respectively, on HAp-coated PLGA/PLA scaffolds. Subcutaneous implantation of the HAp-based 3-D co-culture model into the dorsal region of 6-week-old athymic female nude mice showed better bone-like tissue formation as compared to monoculture of iPSCs-MSCs.

Furthermore, once a tissue-engineered construct is implanted at the defect site, delivery of nutrients and oxygen poses a major challenge in success of a graft. Under in vivo conditions, it is the responsibility of the blood vessels to take care of exchange of nutrients and waste materials. Any tissue that is within 100–200 $\mu$m vasculatures would receive nutrients [145]. Therefore, establishment of pre-vascularized construct is the need of the hour and indeed is an important challenge [146]. In an interesting study, a 3-D co-culture system was established using DPSCs cultured within microcarriers and endothelial cells embedded in type I collagen. Co-cultured constructs demonstrated higher expression of osteogenic markers as compared to monocultures, thereby showing the potential of this system in vascularized bone tissue engineering [147]. In another study, Nguyen et al. designed an in vitro co-culture system to simultaneously culture hMSCs with endothelial cells by encapsulating them in collagen and/or alginate hydrogels and concluded that simultaneous co-culture on collagen hydrogel led to superior outcomes which were further augmented in tubular perfusion system (TPS) bioreactor [148]. Thus, development of such dynamic platforms for pre-angiogenesis is beneficial and can be induced in the construct before in vivo transplantation towards better clinical outcomes. More details on dynamic bone tissue engineering are explained in the later sections.

## 2.5 Role of Growth Factors in Bone Tissue Engineering

Growth factors are endogenously produced large polypeptides and induce various cellular functions such as cell recruitment, their proliferation, migration and differentiation [149]. They also help in formation and maintenance of the newly formed bone tissue [150–152]. During bone fracture healing, associated signalling cascade can be broadly classified into inflammatory factors, angiogenic factors and osteogenic factors [153].

**Inflammatory factors:** Inflammation is the initial stage of bone fracture repair. This phase of healing comprises recruitment of inflammatory cells at the bone fracture site by pro-inflammatory signals, released by platelets [154]. Inflammatory cytokines trigger invasion by lymphocytes, plasma cells, macrophages and osteoclasts. Factors that play a key role during inflammation include interleukin-1 (IL-1), interleukin-6 (IL-6), tumour necrosis factor alpha (TNF-$\alpha$), fibroblast growth factor 2 (FGF-2) and macrophage colony-stimulating factor (M-CSF) [155, 156].

**Angiogenic factors**: Vascularization plays a prominent role in providing oxygen, nutrients and regulatory factors apart from recruiting additional osteoblasts and promotes cell differentiation and endochondral ossification. Impaired or lack of blood supply during bone growth or repair results in tissue hypoxia, bone loss and ultimately in necrosis [157–159]. Hence, impaired blood supply is a major factor in reduced bone healing. Some key angiogenic factors involved in bone regeneration include platelet-derived growth factor (PDGF), bone morphogenetic proteins (BMPs), FGF and vascular endothelial growth factor (VEGF).

**Osteogenic factors**: Bone regeneration calls for recruitment of osteogenic progenitor cells which could differentiate into bone-forming osteoblasts. During ossification, osteoblasts secrete ECM molecules like glycosaminoglycans which interact with growth factors for modulating downstream signalling cascade [160]. Growth factors responsible for triggering the differentiation of progenitor cells into osteogenic lineages are PDGF, TGF-$\beta$, insulin-like growth factors (IGFs), FGF and BMPs. Of these, BMPs are the most widely studied for bone regeneration; BMP-2 and BMP-7 are FDA approved and used clinically for bone regeneration [161, 162].

In summary, in order to achieve effective bone formation, tissue engineering utilizes combination of progenitor cells on bioengineered constructs along with these bioactive molecules.

### 2.5.1   Delivery of Growth Factors

Considering the importance of growth factors in tissue regeneration, their delivery to the site of genesis is of utmost importance. Typically, growth factors do not act in an endocrine fashion; instead, they exhibit short-range diffusion through the ECM and possess short half-lives. Earlier studies involved introduction of growth factors either by direct injection or by systemic local administration. However, these led to suboptimal functioning of the growth factors since these biomolecules have a short half-life due to their rapid degradation in vivo. As a result, in order to maintain supraphysiological concentrations of certain growth factors such as BMP-2, high dose of recombinant BMP-2 is administered. This, however, is associated with cancer development in case of lumbar spinal arthrodesis [163]. Therefore, it becomes imperative to design delivery systems for sustained and controlled release of growth factors in order to reduce multiple administration cycles and associated clinical risks [164]. Depending on the site of delivery and biological requirement, an efficient growth factor carrier must have control over growth factor release kinetics which may be extended, multifactorial or sequential release [165]. As an example, delivery of BMP-2 requires initial burst release followed by slow and gradual release in order to improve bone regeneration [166]. Additionally, the delivery system must be capable of delivering physiologically relevant doses in absolutely targeted fashion while preserving the bioactivity of growth factors for prolonged time periods.

For successful delivery of the right dose and right type of growth factors, various strategies have been employed. These include physical entrapment of growth factors with the scaffold, covalent binding of growth factors to scaffold, affinity-based entrapment and growth factors incorporated within nanocarriers. Each of these has been discussed in the following subsections and depicted as a schematic in Fig. 3.

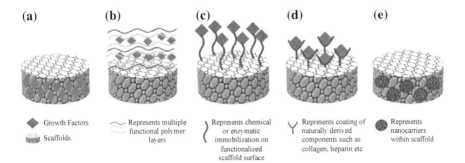

**Fig. 3**  Schematic depicting growth factor immobilization strategies in bone tissue engineering; **a** physical entrapment; **b** layer-by-layer approach; **c** covalent immobilization; **d** affinity-based binding; **e** growth factor-loaded nanocarriers entrapped within scaffolds

Physical Entrapment

In past few decades, physical immobilization of growth factors was limited to their adsorption over the scaffold surface. However, this was associated with poor delivery as described in a study by Ziegler et al. [167], wherein the authors immobilized BMP-2 and basic fibroblast growth factor (bFBF) directly on synthetic bone implants and found that the growth factors lost their biological activity after initial burst period. This drawback strengthened the requirement of immobilizing or entrapping growth factors within the polymeric carriers in order to obtain sustained release with improved biological activity. Thereafter, researchers encapsulated the growth factors within the 3-D constructs by blending them with the carrier polymers prior to fabrication. Physical entrapment of growth factors within the carriers usually does not affect the bioactivity of encapsulated growth factors or scaffold properties.

In a study, Murphy et al. [168] reported the release of VEGF from mineralized, porous PLGA-based scaffolds; VEGF was incorporated during gas foaming/particulate leaching process, and the study showed that VEGF activity was over 70% up to a period of 12 days and had no effect on porosity of the scaffolds. Growth factors can also be entrapped within microspheres to further delay their release. In one such report, Reyes et al. developed a brushite–PLGA system to study the release rates of integrated PDGF, TGF-$\beta$ and VEGF with respect to bone regeneration. The system effectively controlled the release kinetics and maintained the concentration of growth factors at the site of defect, thereby promoting enhanced osteogenesis [169].

Direct adsorption also demonstrated difficulties in controlling the release rates of multiple growth factors. Therefore, researchers came up with a layer-by-layer (LbL) approach, wherein sequential deposition of multiple templates of various synthetic and natural polymers along with bioactive molecules is performed and can be employed for spatial and temporal release of growth factors [170, 171]. The strategy exploits electrostatic interaction between charged substrates and growth factors for the deposition of multiple functional polymer layers over a template [172]. The first LbL film capable of microgram-scale release of BMP-2 was developed by Macdonald et al. [173]. Briefly, they developed a tetralayer architecture using poly (β-aminoester), chondroitin sulfate and BMP-2. In vitro release kinetic studies demonstrated around 10 μg BMP-2 release over a period of two weeks, with less than 1% release in first 3 h when compared to the plain collagen matrices which delivered about a phantom of BMP-2 far quickly to stimulate osteoinduction. The system also retained its biological activity and induced bone differentiation in MC3T3 E1S4 pre-osteoblasts. This method can also be employed to deliver and tune the release of multiple growth factors within the LbL architecture [174]. The aforementioned studies demonstrate potential of LbL technique in precise control over release of multiple growth factors by using polymers with different degradation rates.

Furthermore, rate of release of entrapped growth factors is majorly governed by the polymer property, cross-linking and geometry of the carrier device and is both a diffusion- and degradation-dependent process [175, 176]. Although physical entrapment does not hinder the bioactivity of encapsulated growth factors, only those constructs that do not utilize harsh conditions during scaffold fabrication can be utilized. Further, the loading efficiency is small; only a fraction of growth factors can be bound and have previously demonstrated unpredictable release kinetics. Release of growth factor is also dependent upon the degradation of encapsulating polymer. Thus, in order to control the release kinetics, polymers are functionalized so as to maintain sustained delivery [177].

Covalent Immobilization

This type of approach is generally adapted to promote systemic and prolonged release of growth factors to the cells [178]. Typically, the growth factors are bound to functionalized surfaces by chemical or enzymatic reactions, and their release is mediated by hydrolysis or enzymatic cleavage. Recently, Luca et al. [179] studied the effect of covalently linked BMP-2 and TGF-$\beta$3 on additively manufactured 3-D scaffolds which were modified with poly(oligo (ethylene glycol) methacrylate) brushes for application in osteochondral tissue regeneration. There was significant upregulation of osteochondral differentiation when the growth factors were homogenously linked to the substrate as compared to simple addition of growth factors in soluble form. However, they did not observe any effects when the growth factors were added in a gradient fashion suggesting further optimization of the system.

One of the most popular methods for covalent coupling is through carbodiimide coupling [180]. This has been utilized by Karageorgiou et al. [181] for covalent immobilization of BMP-2 on silk fibroin films. The authors showed that human BMSCs differentiated into osteoblasts when cultured on BMP-2 coupled silk fibroin films in comparison to unmodified silk films in presence of osteogenic stimulants, thereby concluding that the entrapped BMP-2 was more efficient than its delivery in soluble form.

It has also been demonstrated that dopamine can be easily introduced over any organic or inorganic material forming a polydopamine layer which is structurally very similar to 3, 4-dihydroxy-L-phenylalanine [182]. Using this idea, in a study, BMP-2 was loaded onto polydopamine-coated multichannel biphasic calcium phosphate granule system (PD-MCG), and the system showed sustained release of BMP-2 for 30 days. Pre-osteoblast MC3T3-E1 cells seeded on the dopamine-coated biphasic calcium phosphate system displayed enhanced differentiation, and in vivo implantation showed superior bone formation when compared with MCG system without dopamine coating and dopamine-coated MCG void of growth factor. The results demonstrated that PD-MCG could be used as an effective injectable bone substitute to promote new bone formation than those without dopamine coating [183].

While covalent conjugation has been widely explored for growth factor delivery, it can be quite labour intensive and may interfere with the active site of the protein leading to reduced bioactivity. Conjugated growth factors also demonstrate limited diffusion and hence are available only to proximally close cells.

Affinity-Based Binding

Inspired by ECM–growth factor interactions, another pattern for encapsulating growth factors within scaffolds is through the introduction of naturally derived components such as collagen, fibronectin, gelatin and hyaluronic acid as they provide specific biological site for immobilization of growth factors [184–188]. As an example, heparinized scaffolds explore the natural affinity of heparin sulphate and various growth factors such as FGF1, FGF2, VEGF and BMPs [189–191]. In one such example, Kim et al. functionalized PCL/PLGA scaffolds with heparin–dopamine conjugate followed by sequential coating with BMP-2 towards the investigation of osteoblast activity in vitro and bone formation in vivo [192]. It was found that BMP-2-loaded heparin–dopamine-functionalized scaffolds (BMP-2/Hep-DOPA/PCL/PLGA) showed enhanced ALP activity and calcium deposition with osteoblast-like cells in vitro as compared to BMP-2/Hep/PCL/PLGA and PCL/PLGA scaffolds. These results corroborated with better bone formation in vivo as well. In another interesting study by Martino et al. [193], delivery of PDGF-BB and BMP-2 from adhesive fibronectin fragment-functionalized fibrin matrix in a critical-sized calvarial model led to enhanced bone formation at low doses as compared to fibrin-only, fibronectin-functionalized fibrin matrix and fibrin matrix with growth factors.

This type of growth factor encapsulation mechanism protects growth factors from proteolytic degradation and helps them to maintain prolonged biological activity [194, 195]. Such systems are advantageous since there are no chemical treatments done during incorporation of growth factors as the modification of the scaffold is done prior to encapsulation.

Nanocarriers

Nanocarriers have become popular for growth factor delivery due to their high drug loading and retention capacity, large surface area and ability to protect encapsulated protein from in vivo enzymatic degradation [196]. Nanocarriers may be based on synthetic polymers, proteins, polysaccharides, lipids, silica or even nanocapsules. These can be synthesized using various methods like emulsion–solvent evaporation, phase separation, solvent displacement, self-assembly and electrospraying [197]. There are several studies which have reported the effectiveness of nanoparticles in growth factor delivery in context to bone regeneration [198, 199]. In a recent study, osteoinductive and compressive strength of 3-D HAp-based scaffolds was improvised by incorporating BMP-2-loaded nanoparticles. BMP-2-encapsulated PLGA nanoparticles were prepared by double emulsion–solvent evaporation method and were uniformly distributed on the scaffolds using PCL coating. The modified scaffold demonstrated improved bone regeneration capacity in rabbit calvarial bone defect model as compared to uncoated scaffolds.

## *2.6  Dynamic Environment for Bone Tissue Engineering*

Cells in our body reside in a constant flux including the exchange of nutrients and gases (oxygen and carbon dioxide) between cells and the surrounding interstitial fluid. In bone tissue, the shear stress generated by the flow of interstitial fluid provides a mechanical stimulus. This interstitial fluid flow is generated through mechanical loading during movement and locomotion. It has also been demonstrated that longer periods of rest or inactiveness negatively affect bone formation and remodelling [200]. In vitro studies have determined that bone cells respond to the stimulus caused by fluid shear stress by releasing osteogenic factors such as prostaglandins and nitric oxide indicating the relevance of shear stimulus on bone formation [201].

Bone tissue engineering has been previously been reported by many researchers, wherein both static and dynamic conditions have been explored [202]. Static culture is usually performed in a suitable growth medium without any external stimulus or change in dynamics of the system. Under such culture conditions, transport of nutrients and oxygen happens through diffusion, and a diffusion gradient is generated since the cells at surface consume nutrients and oxygen at a faster rate as compared to their supply. Therefore, owing to mass transport limitations at the core of the scaffold, the cells present at the inner core regions do not receive sufficient amount of nutrients and oxygen, thereby leading to cell death and formation of a necrotic core [203]. As a result, a dynamic environment of mass transport and waste removal is not formed, and a physical stimulus similar to the flow of interstitial fluid during in vivo conditions is not present in static atmosphere. These limitations associated with static conditions do not provide a comprehensive and reliable analysis of cell behaviour on the scaffolds through in vitro analysis; therefore, these results cannot be entrusted for advanced analyses such as pre-clinical studies. Such limitations can be overcome by the use of dynamic environments for the cell culture on 3-D scaffolds and can assist in proving realistic answers to the drawbacks associated with the seamless upscaling of in vitro studies to in vivo animal models and thereafter to clinical applications. A comparison of static and dynamic culture is depicted in Table 2. A dynamic environment can either be created in vitro through the use of bioreactors system or by harnessing the in vivo environment to act as a bioreactor for the development of tissue-engineered grafts. Upcoming sections of this chapter will discuss the recent advances in the area of in vitro and in vivo bioreactors.

### 2.6.1  In Vitro Bioreactors

Bioreactors have been used in vitro for expansion of cells on a biomaterial scaffold before in vivo implantation. They allow in vitro culture of cells in a dynamic environment which can be monitored to ensure maximum cell growth under the influence of shear stimulus as well as uniform supply of nutrients and oxygen throughout the scaffold along with the removal of waste products. It has been experimentally demonstrated that shear stress affects the differentiation of MSCs into osteoblasts

**Table 2** Comparison between static and dynamic culture

| Property | Static culture | Dynamic culture |
| --- | --- | --- |
| Growth | Non-uniform cell growth | Uniform cell growth |
| Mass transport | Mass transport purely through diffusion | Mass transport through flow currents generated via shear stress |
| Monitoring of culture conditions | Monitoring cannot be performed | Such system can be monitored for flow velocity, inflow of growth medium and outflow of waste |
| Closeness to in vivo environment | Higher disparity from in vivo environment | Closely mimic in vivo environment |
| Scalability | Does not provide scalable results for follow-up in vivo studies | Provide scalable results for follow-up in vivo studies |

[204, 205]. Shear stress is mainly caused by the flow of liquid medium and can be manipulated by the change in flow velocity, wherein the flow velocity is directly proportional to the shear stress on the cells. Three kinds of flow patterns have been analysed for shear stress including pulsatile, oscillatory and continuous flow. Of these, continuous flow pattern is most utilized for bioreactor studies. The fluid flow velocity and shear stress can be determined by optical measuring techniques such as particle image velocimetry or through computational fluid dynamic (CFD) modelling like finite element analysis and Lattice–Boltzmann method [206, 207]. There are multiple kinds of bioreactors based on the technique applied to create a fluid flow across and around the 3-D scaffolds. Few of these bioreactor types are discussed below and represented in Fig. 4d [208].

Spinner Flask Bioreactors

In order to overcome the limitations associated with static culture, spinner flask bioreactors were introduced to cause a convective flow of media through hydrodynamic forces generated by the spinning of a magnetic stirrer rod placed at the bottom of a cylindrical flask having side arms with filter cap for removal/addition of media or cells. The scaffold is attached at a fixed position through threaded needles which in turn in connected to the top of the container [209, 210] as shown in Fig. 4d. The shear stress caused by convective flow depends on the stirring speed; a study used 30 rotations per minute (rpm) as the stirring speed of the medium in a 120-ml flask, and it was observed that spinner flask culture showed 60% enhanced proliferation at first week and 2.4 times higher ALP activity at 2 weeks in comparison to static culture [211].

In another study, speed of 50 rpm depicted a positive response on osteogenic activity [212]. These are the simplest form of bioreactors, and several studies have

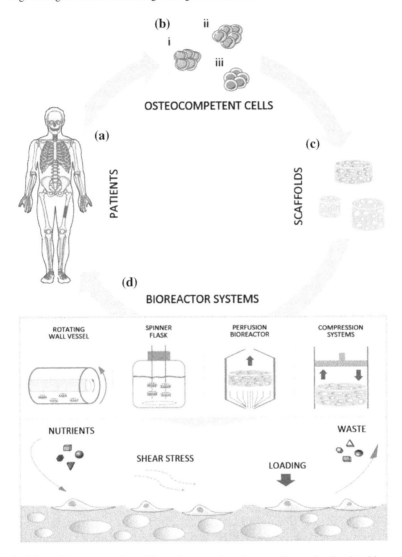

**Fig. 4** Schematic representation of bone tissue engineering paradigm using in vitro bioreactors: **a** patient having a bone defect shown in red colour; **b** osteogenic cells having patient specificity and derived from sources like (i) adult tissue, (ii) induced pluripotent stem cells or (iii) blastocysts generated via somatic cell nuclear transfer; **c** cell-seeded biomaterials scaffolds of different dimensions and porous architecture; **d** dynamic culture of cells on the porous biomaterial scaffolds in different types of bioreactor set-up for the development of bone tissue-engineered grafts. These include **rotating wall vessels**, **spinner flask bioreactors**, **perfusion bioreactors and compression stimulation-based bioreactors.** The function of these bioreactors is to ensure efficient mass transport of nutrients as well as oxygen and to ensure functional regeneration under the effect of mechanically stimulated environment

confirmed the utility of such systems in comparison to static culture for bone formation. Another interesting study by Kim et al. [213] performed on porous silk scaffolds
demonstrated elevated levels of ALP and mineralized matrix along with increased
cell proliferation during dynamic culture in spinner flask as compared to static culture. Despite being useful for dynamic culture of cells, spinner flask culture may be
limited by non-homogeneity of hydrodynamic forces due to high to low gradient of
the flow effect from bottom to top of the flask. Another limitation includes formation
of a dense cell layer on the outer surface of the scaffold [214].

Rotating Wall Vessel (RWV) Bioreactors

RWV bioreactors are primarily composed of a cylindrical vessel that rotates along a
central axis, and this rotational movement produces low shear stress that is sufficient
enough to positively affect cellular activities including proliferation and differentiation. As shown in Fig. 4d, the scaffold inside such rotating chambers can be either
present freely, attached to the vessel wall or present as circular discs along the hori-
zontal axis.
    Numerous studies have shown that RWV bioreactors show favourable response
towards bone formation. A study by Song et al. [215] reported that RWV bioreactors
with fixed scaffold set-up (scaffolds attached on vessel wall) showed five times
higher cell expansion in comparison to stirrer flask bioreactor and static culture. The
cells also showed significantly higher mineralized nodule formation, collagen fibres
and neo-osteoid tissue formation with respect to stirrer flask bioreactor and static
culture. However, in certain studies, lower values for osteogenic markers were found
as compared to spinner flask bioreactor or static culture [216]. This may be attributed
to either the collision of free floating scaffolds or due to lower values of shear stress
causing insignificant stimulus to the cells.

Perfusion Bioreactors

An improved method of dynamic cell culture using hydrodynamic shear stress was
developed in the form of perfusion bioreactor [217]. Instead of indirectly applying
shear forces towards the scaffold through spinning motion (spinner flask bioreactor)
or through rotation (RWV bioreactor), perfusion system directly applies a laminar
fluid flow through the scaffold which enables efficient mass transport of nutrients
and oxygen throughout the scaffold. As shown in Fig. 4d, a perfusion bioreactor
set-up comprises a closed loop of media flow through the scaffold via the assistance
of a peristaltic pump. Inlet and outlet ports are also connected to this system for
media replenishment and waste removal, respectively. There can be two kinds of
flow-perfusion bioreactor systems, indirect and direct perfusion systems. In indirect
perfusion, the scaffold inside the cassette is not sealed tightly, allowing most of
the media fluid to pass through the path of least resistance, i.e. around the scaffold,
although some fluid also passes through the scaffold. Thus, shear stress caused by the

fluid flow is not able to reach the core regions of the scaffold although studies have shown that such systems have been able to provide a favourable osteogenic response as depicted by increase in ALP and osteocalcin protein expression levels [218]. In direct perfusion, scaffold is placed in another cassette that is tightly sealed inside the chamber in a press-fit manner, thus forcing the fluid inflow through the scaffold [219]. The mass transport can easily happen inside the direct perfusion bioreactors, therefore providing better results in comparison to indirect perfusion. Such perfusion bioreactor systems can assist in the development of large tissue-engineered scaffolds for bone applications without formation of a necrotic core [220].

## Stimulation-Based Bioreactor Systems

A dynamic set-up wherein stimulation to osteogenic cells is applied through forces such as mechanical, electromagnetic or ultrasonic stimulus can be considered as stimulation-based bioreactor systems. These are discussed as follows.

### Compression-Based Bioreactors

Mechanical loading and unloading can be sensed by the osteoblasts and osteogenic cells through a phenomenon called mechanosensing [221]. Therefore, scaffold deformation and relaxation through mechanical loading and unloading are sensed by the cells, indirectly via extracellular matrix or directly by change in cell–cell distance or cell shape. This may further modify the local cell environment via change in extracellular gradients or concentration of secreted ligands/growth factors. The compression bioreactor set-up mainly comprises of a cell-seeded scaffold present below a loading piston in which unidirectional load can be applied as static or dynamic compression [222] as shown in Fig. 4d. Several studies have demonstrated the importance of compression loading for bone formation/regeneration. As an example, Matziolis et al. [223] demonstrated that even short-term mechanical stimulation, involving a 24-h cyclic load of 4 kPa and 25% strain at 0.05 Hz, was sufficient to enhance bone formation during culture of human BMSCs on human cancellous bone–fibrin composites inside a compression bioreactor. Another study was performed on biodegradable cryogels composed of L-lactide and dextran with 2-hydroxyethyl methacrylate end groups. These cryogels were seeded with human osteoblast-like cell line, and studies were performed in compression bioreactor with mechanical stimulation at 1.5% strain and at a frequency of 1 Hz for 1 h/day. This study also suggested a positive role of mechanical stimulation in osteogenesis [224].

### Electromagnetic Field (EMF)-Based Bioreactors

Electric and magnetic stimuli have been shown to assist in bone formation and fracture healing since decades [225]. Pulsed electromagnetic fields (PEMF) have been utilized in multiple cases of reduction in bone loss following a fracture [226]. In our body, EMFs and PEMFs are generated during muscle movement. EMFs affect different cellular pathways involved in proliferation and differentiation of cells [227]. The EMF bioreactor set-up for bone tissue engineering applications is essentially composed of Helmholtz coils and PEMF power generator. The cell-adhered scaffolds are placed between the two coils inside a sealed chamber, and electromagnetic field of required intensity is applied. It has been demonstrated through various studies that EMFs induce and promote osteogenesis in MSCs [228] and osteoblasts [229], respectively. In a representative study, Fassina et al. [230] used an electromagnetic bioreactor with a stimulation regime of $2 \pm 0.2$ mT intensity of magnetic field and an electric tension of amplitude $= 5 \pm 1$ mV, frequency $= 75 \pm 2$ Hz and pulse duration $= 1.3$ ms. The Saos-2 human osteoblasts were cultured in porous polyurethane scaffold under the EMF stimulation. Encouraging results were obtained in form of higher cell proliferation and increased coating with decorin and type I collagen along with increased calcium deposition. However, the major disadvantage associated with such system is high cost for set-up, although their advantage lies in the ease of handling due to non-invasiveness and higher possibilities of good manufacturing practice (GMP) approval.

*Combined Stimulation-Based Bioreactors*

In certain bioreactors, two or more types of stimulation methods are applied to enhance the osteogenic response of cells. Some examples of such combined stimulation-based bioreactors include perfusion–compression bioreactor system [231, 232] and EMF-compression-ultrasonic bioreactor system [233]. The cumulative effect of different kind of stimulation increases the osteogenic response as compared to individual stimulation effects due to closer mimicking of native environment of the body, wherein several factors work together to stimulate the cells in order to enhance osteogenesis.

### 2.6.2 In Vivo Bioreactors

In comparison to static methods of culture, in vitro bioreactors are one step closer to the final goal, viz. regeneration of human bone. These bioreactors provide important cues that assist in mimicking the native bone environment; although they can still not match the native environment containing a plethora of pathways and mechanisms working towards cell fate processes. Therefore, in vivo bioreactors have been developed as a recent technology in an effort to provide native environment to the cell–scaffold system, wherein the body of the organism acts as a bioreactor for efficient bone formation using body's own reparative capability [234, 235]. The beauty of such bioreactor system lies in minimal dependency of the cell/scaffold construct on exogenous growth factors, stimulation factors or media supplements since all these requirements can easily be obtained locally from the body of the organism. The term in vivo bioreactors was first introduced in the year 2005 independently by two different research groups. Depending on the model organism being used, studies performed using in vivo bioreactors can be categorized to three types: (a) small animal models, (b) large animal models and (c) clinical studies. These are further discussed as follows:

Small Animal Models

The pioneering studies in the area of in vivo bioreactors were performed in small animal models such as mice, rat and rabbit [236]. In a study by Stevens et al. [237], they demonstrated that an engineered bone having biomechanical properties identical to that of the native bone could be developed by using a "bioreactor" space created in vivo between the tibia and periosteum of New Zealand White rabbits. The engineered bone was harvested after 6 weeks, and thereafter, it was implanted as autologous bone transplantation in contralateral tibial defects leading to complete integration. Another successful study was performed by Holt et al. [238] in a rat model, wherein neovascular in growth and bone formation was achieved by ligating the superficial inferior epigastric vessels through a cylindrical coralline scaffold. In another study, pedicled periosteal flap was utilized as an in vivo bioreactor for development of a vascularized bone graft in a rabbit model [239]. In another in vivo bioreactor study in a rabbit model, tibial periosteum capsule was loaded with 3-D-printed PLA-HAp composite scaffolds. Bone marrow stromal cells were seeded on these scaffolds, and cell-seeded scaffolds were further connected through a vascular supply as illustrated in Fig. 5 [240]. Although these models paved a way for in vivo bioreactors as a new strategy, the scaffold size was too small to understand the complications of large bone implants as applicable under clinical set-up. As a result, large animal models were sought after, as a strategy for advanced studies on in vivo bioreactors.

Large Animal Models

As mentioned previously, large animals like minipigs, sheep, and non-human primate models were used for in vivo bioreactor studies in order to avoid the limitations associated with small animal models. Though non-human primates have many advantages that make them the best models for in vivo bioreactor studies, the associated costs and regulatory concerns limit their usage for broader applications. Akar et al. [241] have listed the prerequisites involved with the use of large animals as the model for in vivo bioreactor studies. As per this list "an in vivo bioreactor should: (1) mimic the clinical surgery techniques; (2) allow evaluation of vascularized bone formation of large volume and complex shape; (3) have an implantation site with high regenerative capacity and low infection risk; (4) be adaptable for different tissue engineering components; (5) allow quantitative evaluation of results; and (6) be available/adaptable in a wide range of clinical research centres". It has been evaluated that ovine (sheep) and porcine (pig) periosteum-guided models fulfil these criterion and therefore have been studied as large animal models for in vivo bioreactor

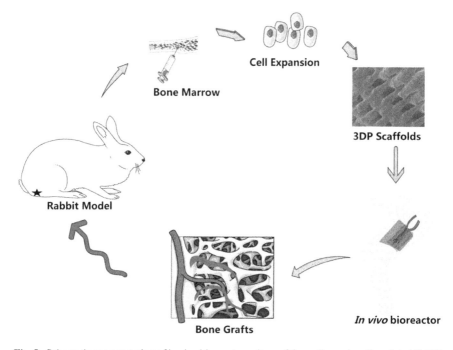

**Fig. 5** Schematic representation of in vivo bioreactor culture of three-dimensionally printed (3-DP) polylactic acid-hydroxyapatite composite scaffolds in a rabbit model. Bone marrow stromal cells were first expanded and then seeded on the 3-DP scaffolds followed by in vivo bioreactor culture in tibial periosteum while being connected to a vascular supply. Such an in vivo bioreactor set-up leads to the formation of vascularized bone tissue

applications. As an example, Cheng et al. [242] observed active bone formation leading to efficient vascularized graft development in an ovine model. In this study, poly (methyl methacrylate) chambers were implanted around rib periosteum. Both muscle fascia and periosteum have also been used as an in vivo bioreactor site, though it has been observed that periosteum serves as a better site in comparison to muscle fascia [243]. In another study, a sheep model implanted with particulate autologous bone graft was studied for development of bone segments via in vivo implantation at the site of rib periosteum [244]. Thus, large animal models help in the understanding the procedure for bone graft development using in vivo bioreactors with the final aim for clinical translation of the strategy.

Clinical Studies

Human body has been used as an in vivo bioreactor in certain studies although such applications are limited in numbers. Warnke et al. [245] used human body as a bioreactor to grow bone on a titanium mesh cage containing bone mineral blocks and BMP-7 along with bone marrow from the patient. These constructs were first implanted at the in vivo bioreactor site for 7 weeks, i.e. latissimus dorsi muscle followed by transplantation for repair of mandibular defects in the same patient. The outcome was favourable as the patient's mastication was improved along with the aesthetics. In another study, a similar scaffold construct was implanted at gastrocolic omentum as the in vivo bioreactor site, before transplantation in mandibular defects. Encouraging outcome was observed in form of better mastication, speech and aesthetics [246]. Some of the points to be taken into consideration for future clinical studies are: (1) evaluating the timing for removal of bone graft from the in vivo bioreactor site or the timing of transplantation at the defect site and (2) involving techniques like microcomputed tomography (micro-CT) and computer-aided design (CAD) to develop custom-designed patient-specific scaffold grafts [247].

## 2.7  Conclusion and Future Perspective

Bone tissue engineering has experienced significant advancements in the recent past, especially in the area of scaffold fabrication, growth factor delivery strategies, usage of cell type as well as bioreactors. More specifically, scaffold fabrication for bone tissue engineering has experienced efforts in the kinds of polymers/metals/ceramics, scaffold architecture as well as surface functionalization. Efforts have also been taken in the direction of growth factor incorporation and cell types utilized. Furthermore, there have been some very interesting reports on the application of dynamic environments (in vitro and in vivo bioreactors). However, very few of these products have reached the clinic. This could be attributed to the uncertainties associated with recapitulation of complex bone environments (physical, mechanical as well as biological properties) within the 3-D matrix. Another critical challenge is the development of a fully vascularized bone graft. This can be achieved by either fabricating scaffolds with large, interconnected pores and also by incorporating angiogenic growth factors. An ideal scaffold that can make a mark in the clinic will be a patient-specific bone graft. More specifically, techniques like 3-D printing may be utilized to print a patient-defect site-specific biomaterial embedded with progenitor cells. The biomaterial could then be subjected to bioreactor-based culture followed by implantation of the patient-specific graft at the defect site.

## References

1. Florencio-silva R, Rodrigues G, Sasso-cerri E (2015) Biology of bone tissue : structure, function, and factors that influence bone cells. Biomed Res Int. https://doi.org/10.1155/2015/421746
2. Weiner S, Wagner HD (1998) The material bone: structure-mechanical function relations. Annu Rev Mater Sci 28:271–298
3. Gordon JAR, Tye CE, Sampaio AV (2007) Bone sialoprotein expression enhances osteoblast differentiation and matrix mineralization in vitro. J Bone 41:462–473

4. Kozhemyakina E, Lassar AB, Zelzer E (2015) A pathway to bone: signaling molecules and transcription factors involved in chondrocyte development and maturation. Development 142:817–831

5. Allen MR, Burr DB (2014) Bone modeling and remodeling. In: Burr DB, Allen MRBT-B ABB (eds) Academic Press, San Diego, pp 75–90

6. Gruber R, Koch H, Doll BA (2006) Fracture healing in the elderly patient. Exp Gerontol 41:1080–1093

7. Gandhi A, Liporace F, Azad V (2006) Diabetic fracture healing. Foot Ankle Clin 11:805–824

8. Wukich DK, Kline AJ (2008) The management of ankle fractures in patients with diabetes. J Bone Jt Surg–Am 90:1570–1578

9. Lu C, Hansen E, Sapozhnikova A (2010) Effect of age on vascularization during fracture repair. J Orthop Res 26:1384–1389

10. Mobini S, Ayoub A (2016) Bone tissue engineering in the maxillofacial region: the state-of-the-art practice and future prospects. Regen Reconstr Restor 1:8–14

11. Amini AR, Laurencin CT, Nukavarapu SP (2012) Bone tissue engineering: recent advances and challenges. Crit Rev Biomed Eng 40:363–408

12. Delloye C, Cornu O, Druez V, Barbier O (2007) Bone allografts: what they can offer and what they cannot. J Bone Jt Surg 89–B:5:574–580

13. Yeatts AB, Fisher JP (2011) Bone tissue engineering bioreactors: dynamic culture and the influence of shear stress. Bone 48:171–181

14. Della Porta G, Nguyen BB, Campardelli R (2014) Synergistic effect of sustained release of growth factors and dynamic culture on osteoblastic differentiation of mesenchymal stem cells. J Biomed Mater Res Part A 103:2161–2171

15. Nguyen BB, Ko H, Moriarty RA (2016) dynamic bioreactor culture of high volume engineered bone tissue. Tissue Eng Part A 22:263–271

16. Brien FJO (2011) Biomaterials & scaffolds for tissue engineering. Mater Today 14:88–95

17. Hutmacher DW (2000) Scaffolds in tissue engineering bone and cartilage. Biomaterials 21:2529–2543

18. Williams DF (2008) On the mechanisms of biocompatibility. Biomaterials 29:2941–2953

19. Prasadh S, Chung R, Wong W (2018) Unraveling the mechanical strength of biomaterials used as a bone scaffold in oral and maxillofacial defects. Oral Sci Int 15:48–55

20. Zhao H, Liang W (2017) A novel comby scaffold with improved mechanical strength for bone tissue engineering. Mater Lett 194:220–223

21. Chen Y, Frith JE, Dehghan- A (2017) Mechanical properties and biocompatibility of porous titanium scaffolds for bone tissue engineering. J Mech Behav Biomed Mater 75:169–174

22. Li G, Wang L, Pan W (2016) In vitro and in vivo study of additive manufactured porous Ti6Al4V scaffolds for repairing bone defects. Sci Rep 6:1–11

23. Zhang X, Zeng D, Li N (2016) Functionalized mesoporous bioactive glass scaffolds for enhanced bone tissue regeneration. Sci Rep 6:19361

24. Nisal A, Sayyad R, Dhavale P (2018) Silk fibroin micro-particle scaffolds with superior compression modulus and slow bioresorption for effective bone regeneration. Sci Rep 1–10

25. Persson M, Lehenkari PP, Berglin L (2018) Osteogenic differentiation of human mesenchymal stem cells in a 3D woven scaffold. Sci Rep 8:1–12

26. Schieker M, Seitz H, Drosse I (2006) Biomaterials as scaffold for bone tissue engineering. Eur J Trauma 32:114–124

27. Alaribe FN, Manoto SL, Motaung SCKM (2016) Scaffolds from biomaterials: advantages and limitations in bone and tissue engineering. Biologia (Bratisl) 71:353–366

28. Puppi D, Chiellini F, Piras AM, Chiellini E (2010) Progress in polymer science polymeric materials for bone and cartilage repair. Prog Polym Sci 35:403–440

29. Wang Z, Lin M, Xie Q (2016) Electrospun silk fibroin/poly(lactide-co-ε-caprolactone) nanofibrous scaffolds for bone regeneration. Int J Nanomed 11:1483–1500

30. Zhang Y, Reddy VJ, Wong SY (2010) Enhanced biomineralization in osteoblasts on a novel electrospun biocomposite nanofibrous substrate. Tissue Eng Part A 16:1949–1960

31. Jing Z, Wu Y, Su W (2017) carbon nanotube reinforced collagen/hydroxyapatite scaffolds improve bone tissue formation in vitro and in vivo. Ann Biomed Eng 45:2075–2087
32. Chevalier J, Gremillard L (2009) Ceramics for medical applications: a picture for the next 20 years. J Eur Ceram Soc 29:1245–1255
33. Thein-han W, Xu HHK (2011) Collagen-calcium phosphate cement scaffolds seeded with umbilical cord stem cells for bone tissue engineering. Tissue Eng Part A 17:2943–2954
34. Bose S, Roy M, Bandyopadhyay A (2012) Recent advances in bone tissue engineering scaffolds. Trends Biotechnol 30:546–554
35. Asaoka T, Ohtake S, Furukawa KS (2013) Development of bioactive porous a-TCP/HAp beads for bone tissue engineering. J Biomed Mater Res, Part A 101:3295–3300
36. Wu C, Chang J (2007) Degradation, bioactivity, and cytocompatibility of diopside, akermanite, and bredigite ceramics. J Biomed Mater Res Part B Appl Biomater 83B:153–160
37. Fu Q, Saiz E, Rahaman MN, Tomsia AP (2011) Bioactive glass scaffolds for bone tissue engineering: state of the art and future perspectives. Mater Sci Eng, C 31:1245–1256
38. Xia L, Yin Z, Mao L (2016) Akermanite bioceramics promote osteogenesis, angiogenesis and suppress osteoclastogenesis for osteoporotic bone regeneration. Sci Rep 6:1–17
39. Jones JR (2013) Review of bioactive glass: from Hench to hybrids. Acta Biomater 9:4457–4486
40. Philippart A, Boccaccini AR, Fleck C (2015) Toughening and functionalization of bioactive ceramic and glass bone scaffolds by biopolymer coatings and infiltration: a review of the last 5 years. Expert Rev Med Devices 12:93–111
41. Staiger MP, Pietak AM, Huadmai J, Dias G (2006) Magnesium and its alloys as orthopedic biomaterials: a review. Biomaterials 27:1728–1734
42. Alvarez K, Nakajima H (2009) Metallic scaffolds for bone regeneration. Materials (Basel) 2:790–832
43. Hussein MA, Mohammed AS, Al-aqeeli N, Arabia S (2015) Wear characteristics of metallic biomaterials: a review. Materials (Basel) 8:2749–2768
44. Ghassemi T, Shahroodi A, Ebrahimzadeh MH, Mousavian A (2018) Current concepts in scaffolding for bone tissue engineering. Arch Bone Jt Surg 6:90–99
45. Bhui AS, Singh G, Sidhu SS, Bains PS (2018) Experimental investigation of optimal ED machining parameters for Ti-6Al-4V biomaterial. FU Ser Mech Eng 16(3):337–345
46. Tamaddon M, Samizadeh S, Wang L (2017) Intrinsic osteoinductivity of porous titanium scaffold for bone tissue engineering. Int J Biomater 2017:5093063
47. Bobe K, Willbold E, Morgenthal I (2013) In vitro and in vivo evaluation of biodegradable, open-porous scaffolds made of sintered magnesium W4 short fibres. Acta Biomater 9:8611–8623
48. Wu C, Zhou Y, Xu M (2013) Copper-containing mesoporous bioactive glass scaffolds with multifunctional properties of angiogenesis capacity, osteostimulation and antibacterial activity. Biomaterials 34:422–433
49. Declercq HA, Desmet T, Dubruel P, Cornelissen MJ (2014) The role of scaffold architecture and composition on the bone formation by adipose-derived stem cells. Tissue Eng Part A 20:434–444
50. Kim YE, Kim Y (2013) Effect of biopolymers on the characteristics and cytocompatibility of biocomposite nanofibrous scaffolds. Polym J 45:845–853
51. Roohani-Esfahani SI, Newman P, Zreiqat H (2016) Design and fabrication of 3D printed scaffolds with a mechanical strength comparable to cortical bone to repair large bone defects. Sci Repor 1–8
52. Leite ÁJ, Oliveira NM, Song W, Mano JF (2018) Bioactive hydrogel marbles. Sci Rep 8:1–11
53. Sato Y, Yamamoto K, Horiguchi S (2018) Nanogel tectonic porous 3D scaffold for direct reprogramming fibroblasts into osteoblasts and bone regeneration. Sci Rep 8:15824
54. Nicodemus GD, Bryant SJ (2008) Cell encapsulation in biodegradable hydrogels for tissue engineering applications. Tissue Eng Part B 14:149–165
55. Short AR, Koralla D, Deshmukh A (2015) Hydrogels that allow and facilitate bone repair, remodeling, and regeneration. J Mater Chem B 3:7818–7830

56. Tibbitt MW, Anseth KS (2009) Hydrogels as extracellular matrix mimics for 3D cell culture. Biotechnol Bioeng 103:655–663
57. Yang X, Sun T, Dou S (2009) Block copolymer of polyphosphoester and poly (L-Lactic Acid) modified surface for enhancing osteoblast adhesion, proliferation, and function. Biomacromol 10:2213–2220
58. Thambi T, Li Y, Lee DS (2017) Injectable hydrogels for sustained release of therapeutic agents. J Control Release 267:57–66
59. Liu M, Zeng X, Ma C (2017) Injectable hydrogels for cartilage and bone tissue engineering. Bone Res 5:17014
60. Han Y, Zeng Q, Li H, Chang J (2013) The calcium silicate/ alginate composite: preparation and evaluation of its behavior as bioactive injectable hydrogels. Acta Biomater 9:9107–9117
61. Ding C, Zhao L, Liu F (2010) Dually responsive injectable hydrogel prepared by in situ cross-linking of glycol chitosan and benzaldehyde-capped PEO-PPO-PEO. Biomacromol 11:1043–1051
62. Dessı M, Borzacchiello A, Mohamed THA (2013) Novel biomimetic thermosensitive b - tricalcium phosphate/ chitosan-based hydrogels for bone tissue engineering. J Biomed Mater Res - Part A 101:2984–2993
63. Vo TN, Shah SR, Lu S (2016) Injectable dual-gelling cell-laden composite hydrogels for bone tissue engineering. Biomaterials 83:1–11
64. Fu S, Ni P, Wang B (2012) Injectable and thermo-sensitive PEG-PCL-PEG copolymer/collagen/n-HA hydrogel composite for guided bone regeneration. Biomaterials 33:4801–4809
65. Jiao Y, Gyawali D, Stark JM (2012) A rheological study of biodegradable injectable PEGMC/HA composite scaffolds. Soft Matter 8:1499–1507
66. Niranjan R, Koushik C, Saravanan S (2013) A novel injectable temperature-sensitive zinc doped chitosan/beta-glycerophosphate hydrogel for bone tissue engineering. Int J Biol Macromol 54:24–29
67. Douglas TEL, Piwowarczyk W, Pamula E (2014) Injectable self-gelling composites for bone tissue engineering based on gellan gum hydrogel enriched with different bioglasses. Biomed Mater 9:045014
68. Dhivya S, Saravanan S, Sastry TP, Selvamurugan N (2015) Nanohydroxyapatite—reinforced chitosan composite hydrogel for bone tissue repair in vitro and in vivo. J Nanobiotechnology 13:40
69. Lewandowska-ła J, Fiejdasz S, Rodzik Ł (2015) Bioactive hydrogel-nanosilica hybrid materials: a potential injectable scaffold for bone tissue engineering. Biomed Mater 10:015020
70. Huang Y, Zhang X, Wu A, Xu H (2016) Injectable nano-hydroxyapatite (n-HA)/glycol chitosan(G-CS)/hyaluronic acid (HyA) composite hydrogel for bone tissue engineering. RSC Adv 6:33529–33536
71. Tan R, She Z, Wang M (2012) Thermo-sensitive alginate-based injectable hydrogel for tissue engineering. Carbohydr Polym 87:1515–1521
72. Celikkin N, Mastrogiacomo S, Jaroszewicz J, Walboomers XF (2018) Gelatin methacrylate scaffold for bone tissue engineering: the influence of polymer concentration. J Biomed Mater Res Part A 106A:201–209
73. Ko W, Lee JS, Hwang Y (2018) Injectable hydrogel composite containing modified gold nanoparticles: implication in bone tissue regeneration. Int J Nanomedicine 13:7019–7031
74. Arya N, Sardana V, Saxena M (2012) Recapitulating tumour microenvironment in chitosan-gelatin three-dimensional scaffolds: an improved in vitro tumour model. J R Soc Interface 9:3288–3302
75. Bencherif SA, Braschler TM, Renaud P (2013) Advances in the design of macroporous polymer scaffolds for potential applications in dentistry. J Periodontal Implant Sci 43:251–261
76. Elamparithi D, Moorthy V (2017) On various porous scaffold fabrication methods. Mapana J Sci 16:47–52
77. Tamburaci S, Tihminlioglu F (2018) Materials science & engineering C Biosilica incorporated 3D porous scaffolds for bone tissue engineering applications. Mater Sci Eng, C 91:274–291

78. Demirtas TT, Irmak G, Gümüşderelioglu M (2017) Bioprintable form of chitosan hydrogel for bone tissue engineering. Biofabrication 9:035003
79. Moncal KK, Heo DN, Godzik KP (2018) 3D printing of poly (e-caprolactone)/poly (D, L-lactide-co-glycolide)/hydroxyapatite composite constructs for bone tissue engineering. J Mater Res 33:1972–1986
80. Nandi SK, Fielding G, Banerjee D (2018) 3D-printed b-TCP bone tissue engineering scaffolds: effects of chemistry on in vivo biological properties in a rabbit tibia model. J Mater Res 33:1939–1947
81. Zhang J, Chen Y, Xu J (2018) Tissue engineering using 3D printed nano-bioactive glass loaded with NELL1 gene for repairing alveolar bone defects. Regen Biomater 5:213–220
82. Ma H, Feng C, Chang J, Wu C (2018) 3D-printed bioceramic scaffolds : from bone tissue engineering to tumor therapy. Acta Biomater 1–23
83. Jammalamadaka U, Tappa K (2018) Recent advances in biomaterials for 3D printing and tissue engineering. J Funct Biomater 9:14
84. Karageorgiou V, Kaplan D (2005) Porosity of 3D biomaterial scaffolds and osteogenesis. Biomaterials 26:5474–5491
85. Guda T, Walker JA, Singleton B (2014) Hydroxyapatite scaffold pore architecture effects in large bone defects in vivo. J Biomater Appl 28:1016–1027
86. Gupte MJ, Swanson WB, Hu J (2018) Pore size directs bone marrow stromal cell fate and tissue regeneration in nanofibrous macroporous scaffolds by mediating vascularization. Acta Biomater 82:1–11
87. Xu M, Zhai D, Chang J, Wu C (2014) In vitro assessment of three-dimensionally plotted nagelschmidtite bioceramic scaffolds with varied macropore morphologies. Acta Biomater 10:463–476
88. Smith LA, Liu X, Ma PX (2008) Tissue engineering with nano-fibrous scaffolds. Soft Matter 4:2144–2149
89. Liu H, Ding X, Zhou G (2013) Electrospinning of nanofibers for tissue engineering applications. J Nanomater 2013:495708
90. Lyu S, Huang C, Yang H, Zhang X (2013) Electrospun fibers as a scaffolding platform for bone tissue repair. J Orthop Res 31:1382–1389
91. Guo Z, Xu J, Ding S (2015) In vitro evaluation of random and aligned polycaprolactone/gelatin fibers via eletrospinning for bone tissue engineering. J Biomater Sci Polym Ed 26:989–1001
92. Wang Y, Cai X, Wang Y (2016) Enhanced osteogenesis of BMP2-Transfected human periodontal ligament stem cells by aligned electrospun scaffolds for bone tissue engineering. J Biomater Tissue Eng 6:563–573
93. Chen H, Qian Y, Xia Y (2016) Enhanced osteogenesis of ADSCs by the synergistic effect of aligned fibers containing collagen I. ACS Appl Mater Interfaces 8:29289–29297
94. Stevens MM (2008) Biomaterials for bone tissue engineering. Mater Today 11:18–25
95. Turnbull G, Clarke J, Picard F (2018) 3D bioactive composite scaffolds for bone tissue engineering. Bioact Mater 3:278–314
96. De Witte T, Fratila-apachitei LE, Zadpoor AA, Peppas NA (2018) Bone tissue engineering via growth factor delivery: from scaffolds to complex matrices. Regen Biomater 5:197–211
97. Benoit DSW, Schwartz MP, Durney AR, Anseth KS (2008) Small functional groups for controlled differentiation of hydrogel-encapsulated human mesenchymal stem cells. Nat Mater 7:816–823
98. Arora A, Katti DS (2016) Understanding the influence of phosphorylation and polysialylation of gelatin on mineralization and osteogenic differentiation. Mater Sci Eng, C 65:9–18
99. Jaidev LR, Chatterjee K (2019) Surface functionalization of 3D printed polymer scaffolds to augment stem cell response. Mater Des 161:44–54
100. Yang F, Williams CG, Wang D-A (2005) The effect of incorporating RGD adhesive peptide in polyethylene glycol diacrylate hydrogel on osteogenesis of bone marrow stromal cells. Biomaterials 26:5991–5998
101. Chen W, Zhou H, Weir MD (2012) Human embryonic stem cell-derived mesenchymal stem cell seeding on calcium phosphate cement-chitosan-RGD scaffold for bone repair. Tissue Eng Part A 21201:1–37

102. Beck G, Crichton HJ, Baer E (2014) Surface modifying oligomers used to functionalize polymeric surfaces: consideration of blood contact applications. J Appl Polym Sci 131:40328
103. Gümüsderelioglu M, Aday S (2011) Heparin-functionalized chitosan scaffolds for bone tissue engineering. Carbohydr Res 346:606–613
104. Kim HD, Lee EA, An Y (2017) Chondroitin sulfate-based biomineralizing surface hydrogels for bone tissue engineering. ACS Appl Mater Interfaces 9:21639–21650
105. Downey PA, Siegel MI (2006) Bone biology and the clinical implications for osteoporosis. Phys Ther 86:77–91
106. Cancedda R, Giannoni P, Mastrogiacomo M (2007) A tissue engineering approach to bone repair in large animal models and in clinical practice. Biomaterials 28:4240–4250
107. Heath CA (2000) Cells for tissue engineering. Trends Biotechnol 18:17–19
108. Krampera M, Pizzolo G, Aprili G, Franchini M (2006) Mesenchymal stem cells for bone, cartilage, tendon and skeletal muscle repair. Bone 39:678–683
109. Uccelli A, Moretta L, Pistoia V (2008) Mesenchymal stem cells in health and disease. Nat Rev Immunol 8:726–736
110. Hanson S, Souza RND, Hematti P (2014) Biomaterial-mesenchymal stem cell constructs for immunomodulation in composite tissue engineering. Tissue Eng Part A 20:2162–2168
111. Nassif L, Jurjus A, Nassar J (2012) Enhanced in vivo bone formation by bone marrow differentiated mesenchymal stem cells grown in chitosan scaffold. J Bioeng Biomed Sci 2:2–7
112. Yu L, Wu Y, Liu J (2018) 3D culture of bone marrow-derived mesenchymal stem cells (BMSCs) could improve bone regeneration in 3D-printed porous Ti6Al4V scaffolds. Stem Cells Int 2018:2074021
113. Krampera M, Pasini A, Pizzolo G (2006) Regenerative and immunomodulatory potential of mesenchymal stem cells. Curr Opin Pharmacol 6:435–441
114. Caplan AI (2007) Adult mesenchymal stem cells for tissue engineering versus regenerative medicine. J Cell Physiol 213:341–347
115. Frese L, Dijkman E, Hoerstrup SP (2016) Adipose tissue-derived stem cells in regenerative medicine. Transfus Med Hemotherapy 43:268–274
116. Calabrese G, Giuff R, Forte S (2017) Human adipose-derived mesenchymal stem cells seeded into a collagen-hydroxyapatite scaffold promote bone augmentation after implantation in the mouse. Sci Rep 7:7110
117. Sayin E, Rashid RH, Rodríguez-Cabello JC (2017) Human adipose derived stem cells are superior to human osteoblasts (HOB) in bone tissue engineering on a collagen- fibroin-ELR blend. Bioact Mater 1–11
118. Oryan A, Kamali A, Moshiri A, Eslaminejad MB (2017) Role of mesenchymal stem cells in bone regenerative medicine: what is the evidence? Cells Tissues Organs 204:59–83
119. Yu J, Wang Y, Deng Z (2007) Odontogenic capability: bone marrow stromal stem cells versus dental pulp stem cells. Biol Cell 99:465–474
120. El-gendy R, Yang XB, Newby PJ (2013) Osteogenic differentiation of human dental pulp stromal cells on 45S5 bioglass based scaffolds in vitro and in vivo. Tissue Eng Part A 19:707–715
121. Petridis X, Diamanti E, Trigas GC, Kalyvas D (2015) Bone regeneration in critical-size calvarial defects using human dental pulp cells in an extracellular matrix-based scaffold. J Craniomaxillofac Surg
122. La M, Paino F, Spina A (2014) Dental pulp stem cells: state of the art and suggestions for a true translation of research into therapy. J Dent 42:761–768
123. Tian X, Heng B, Ge Z (2008) Comparison of osteogenesis of human embryonic stem cells within 2D and 3D culture systems. Scand J Clin Lab Investig 68:58–67
124. De Peppo GM, Marcos-campos I, John D (2013) Engineering bone tissue substitutes from human induced pluripotent stem cells. Proc Natl Acad Sci 110:8680–8685
125. Marolt D, Marcos I, Bhumiratana S (2012) Engineering bone tissue from human embryonic stem cells. Proc Natl Acad Sci 109:1–5
126. Vats A, Tolley NS, Bishop AE, Polak JM (2005) Embryonic stem cells and tissue engineering: delivering stem cells to the clinic. J R Soc Med 98:346–350

127. Swijnenburg R, Schrepfer S, Govaert JA (2008) Immunosuppressive therapy mitigates immunological rejection of human embryonic stem cell xenografts. Proc Natl Acad Sci 105:12991–12996

128. Wu Q, Yang B, Hu K (2017) Deriving osteogenic cells from induced pluripotent stem cells for bone tissue engineering. Tissue Eng Part B Rev 23:1–8

129. Jin G, Kim T, Kim J (2012) Bone tissue engineering of induced pluripotent stem cells cultured with macrochanneled polymer scaffold. J Biomed Mater Res Part A 101:1283–1291

130. Hayashi K, Ochiai-shino H, Shiga T (2016) Transplantation of human-induced pluripotent stem cells carried by self-assembling peptide nanofiber hydrogel improves bone regeneration in rat calvarial bone defects. BDJOpen 2:1–7

131. Csobonyeiova M, Polak S, Zamborsky R, Danisovic L (2017) iPS cell technologies and their prospect for bone regeneration and disease modeling: a mini review. J Adv Res 8:321–327

132. Xie J, Peng C, Zhao Q (2015) Osteogenic differentiation and bone regeneration of the iPSC-mscs supported by a biomimetic nanofibrous scaffold. Acta Biomater 29:365–379

133. Jung Y, Bauer G, Nolta JA (2012) Concise review: induced pluripotent stem cell-derived mesenchymal stem cells: progress toward safe clinical products. Stem Cells 30:42–47

134. Reible B, Schmidmaier G, Prokscha M, Westhauser F (2017) Continuous stimulation with differentiation factors is necessary to enhance osteogenic differentiation of human mesenchymal stem cells. Growth Factors 35:179–188

135. Kimelman N, Pelled G, Helm GA (2007) Review: gene- and stem cell-based therapeutics for bone regeneration and repair. Tissue Eng 13:1135–1150

136. Huynh NPT, Brunger JM, Gloss CC (2018) Genetic engineering of mesenchymal stem cells for differential matrix deposition on 3D woven scaffolds. Tissue Eng Part A 00:1–14

137. Kuttappan S, Anitha A, Minsha MG (2018) BMP2 expressing genetically engineered mesenchymal stem cells on composite fibrous scaffolds for enhanced bone regeneration in segmental defects. Mater Sci Eng, C 85:239–248

138. Kargozar S, Hashemian SJ, Soleimani M (2017) Acceleration of bone regeneration in bioactive glass/gelatin composite scaffolds seeded with bone marrow-derived mesenchymal stem cells over-expressing bone morphogenetic protein-7. Mater Sci Eng, C 75:688–698

139. Sayed N, Liu C, Wu JC (2016) Translation of human-induced pluripotent stem cells from clinical trial in a dish to precision medicine. J Am Coll Cardiol 67:2161–2176

140. Martin U (2017) Therapeutic application of pluripotent stem cells: challenges and risks. Front Med 4:229

141. Raggatt LJ, Partridge NC (2010) Cellular and molecular mechanisms of bone remodeling. J Biol Chem 285:25103–25108

142. Kook Y, Jeong Y, Lee K, Koh W (2017) Design of biomimetic cellular scaffolds for co-culture system and their application. J Tissue Eng 8:1–17

143. Beşkardeş IG, Hayden RS, Glettig DL (2017) Bone tissue engineering with scaffold-supported perfusion co-cultures of human stem cell-derived osteoblasts and cell line-derived osteoclasts. Process Biochem 59:303–311

144. Jeon OH, Panicker LM, Lu Q (2016) Human iPSC-derived osteoblasts and osteoclasts together promote bone regeneration in 3D biomaterials. 1–11

145. Lovett M, Ph D, Lee K (2009) Vascularization strategies for tissue engineering. Tissue Eng Part B 15:353–370

146. Gurel G, Torun G, Hasirci V (2016) Influence of co-culture on osteogenesis and angiogenesis of bone marrow mesenchymal stem cells and aortic endothelial cells. Microvasc Res 108:1–9

147. Jin G, Kim H (2017) Co-culture of human dental pulp stem cells and endothelial cells using porous biopolymer microcarriers: a feasibility study for bone tissue engineering. Tissue Eng anf Regen Med 14:393–401

148. Nguyen BB, Moriarty RA, Kamalitdinov T (2018) Collagen hydrogel scaffold promotes mesenchymal stem cell and endothelial cell coculture for bone tissue engineering. J Biomed Mater Res Part A 105:1123–1131

149. Lee K, Silva EA, Mooney DJ (2011) Growth factor delivery-based tissue engineering: general approaches and a review of recent developments. J R Soc Interface 8:153–170

150. Canalis E (2009) Prospect-growth factor control of bone mass. J Cell Biochem 108:769–777
151. Li F, Niyibizi C (2012) Cells derived from murine induced pluripotent stem cells (iPSC) by treatment with members of TGF-beta family give rise to osteoblasts differentiation and form bone in vivo. BMC Cell Biol 13:35
152. Nyberg E, Holmes C, Witham T, Grayson WL (2015) Growth factor-eluting technologies for bone tissue engineering. Drug Deliv Transl Res 6:184–194
153. Vo TN, Kasper FK, Mikos AG (2012) Strategies for controlled delivery of growth factors and cells for bone regeneration. Adv Drug Deliv Rev. https://doi.org/10.1016/j.addr.2012.01.016
154. Hankenson K, Gagne K, Shaughnessy M (2015) Extracellular signaling molecules to promote fracture healing and bone regeneration. Adv Drug Deliv Rev 94:3–12. https://doi.org/10.1016/j.addr.2015.09.008
155. Lienemann PS, Lutolf MP, Ehrbar M (2012) Biomimetic hydrogels for controlled biomolecule delivery to augment bone regeneration. Adv Drug Deliv Rev 64:1078–1089
156. Kowalczewski CJ, Saul JM (2018) Biomaterials for the delivery of growth factors and other therapeutic agents in tissue engineering approaches to bone regeneration. Front Pharmacol 9:513
157. Marenzana M, Arnett TR (2013) The key role of the blood supply to bone. Bone Res 1:203–215
158. Nauta TD, Van Hinsbergh VWM, Koolwijk P (2014) Hypoxic signaling during tissue repair and regenerative medicine. Int J Mol Sci 15:19791–19815
159. Mercado-Pagan AE, Stahl AM, Shanjani Y, Yang Y (2015) Vascularization in bone tissue engineering constructs. Ann Biomed Eng 43:718–729
160. Nikitovic D, Zafiropoulos A, Tzanakakis GN (2005) Effects of glycosaminoglycans on cell proliferation of normal osteoblasts and human osteosarcoma cells depend on their type and fine chemical compositions. Anticancer Res 25:2851–2856
161. Nauth A, Ristevski B, Li R, Schemitsch EH (2011) Growth factors and bone regeneration: how much bone can we expect? Injury 42:574–579
162. Farokhi M, Mottaghitalab F, Shokrgozar MA (2016) Importance of dual delivery systems for bone tissue engineering. J Control Release 225:152–169
163. Carragee EJ, Comer G, Chu G (2013) Cancer risk after use of recombinant bone. J Bone Jt Surg 95:1537–1545
164. Wang W, Yeung KWK (2017) Bioactive materials bone grafts and biomaterials substitutes for bone defect repair: a review. Bioact Mater 2:224–247
165. Fernandez-Yague MA, Abbah SA, McNamara L (2014) Biomimetic approaches in bone tissue engineering: integrating biological and physicomechanical strategies. Adv Drug Deliv Rev 84:1–29
166. Brown KV, Lon M, Li B (2011) Improving bone formation in a rat femur segmental defect by controlling bone morphogenetic protein-2 release. Tissue Eng Part A 17:1735–1746
167. Ziegler J, Anger D, Krummenauer F (2007) Biological activity of recombinant human growth factors released from biocompatible bone implants. https://doi.org/10.1002/jbm.a.31625
168. Murphy WL, Peters MC, Kohn DH, Mooney DJ (2000) Sustained release of vascular endothelial growth factor from mineralized poly (lactide-co-glycolide) scaffolds for tissue engineering. Biomater 21:2521–2527
169. Reyes R, De B, Delgado A (2012) Effect of triple growth factor controlled delivery by a brushite—PLGA system on a bone defect. Injury 43:334–342
170. Wang Y, Angelatos AS, Caruso F (2008) Template synthesis of nanostructured materials via layer-by-layer. Chem Mater 20:848–858
171. Wang Z, Wang Z, Lu WW (2017) Novel biomaterial strategies for controlled growth factor delivery for biomedical applications. NPG Asia Mater 9:e435
172. Richardson JJ, Björnmalm M, Caruso F (2015) Technology-driven layer-by-layer assembly of nanofilms. Science 80(348):411–424
173. Macdonald ML, Samuel RE, Shah NJ (2011) Tissue integration of growth factor-eluting layer-by-layer polyelectrolyte multilayer coated implants. Biomaterials 32:1446–1453
174. Shah NJ, Macdonald ML, Beben YM (2011) Tunable dual growth factor delivery from polyelectrolyte multilayer films. Biomaterials 32:6183–6193

175. Bouyer M, Guillot R, Jonathan L (2016) Surface delivery of tunable doses of BMP-2 from an adaptable polymeric scaffold induces volumetric bone regeneration. Biomaterials 104:168–181
176. Newman MR, Benoit DSW (2016) Local and targeted drug delivery for bone regeneration. Curr Opin Biotechnol 40:125–132
177. Draenert FG, Nonnenmacher A, Ka PW (2012) BMP-2 and bFGF release and in vitro effect on human osteoblasts after adsorption to bone grafts and biomaterials. Clin Oral Implant Res 24:750–757
178. Masters KS (2011) Covalent growth factor immobilization strategies for tissue repair and regeneration. Macromol Biosci 11:1149–1163
179. Di Luca A, Klein-gunnewiek M, Vancso JG (2017) Covalent binding of bone morphogenetic protein-2 and transforming growth factor-β3 to 3D plotted scaffolds for osteochondral tissue regeneration. Biotechnol J 12:1700072
180. Madl CM, Mehta M, Duda GN (2013) Presentation of BMP-2 mimicking peptides in 3D hydrogels directs cell fate commitment in osteoblasts and mesenchymal stem cells. Biomacromol 15:445–455
181. Karageorgiou V, Meinel L, Hofmann S (2004) Bone morphogenetic protein-2 decorated silk fibroin films induce osteogenic differentiation of human bone marrow stromal cells. J Biomed Mater Res Part A 71A:528–537
182. Lee H, Dellatore SM, Miller WM, Messersmith PB (2007) Mussel-inspired surface chemistry for multifunctional coatings. Science (80-) 318:426–431
183. Lee GH, Paul K, Lee B (2017) Development of BMP-2 immobilized polydopamine mediated multichannelled biphasic calcium phosphate granules for improved bone regeneration. Mater Lett 208:122–125
184. Geiger M, Li RH, Friess W (2003) Collagen sponges for bone regeneration with rhBMP-2. Adv Drug Deliv Rev 55:1613–1629
185. Upton Z, Cuttle L, Noble A (2008) Vitronectin: growth factor complexes hold potential as a wound therapy approach. J Invest Dermatol 128:1534–1544
186. Schultz GS, Wysocki A (2009) Interactions between extracellular matrix and growth factors in wound healing. Wound Repair Regen 17:153–162
187. Martino MM, Briquez PS, Ranga A (2013) Heparin-binding domain of fibrin(ogen) binds growth factors and promotes tissue repair when incorporated within a synthetic matrix. Proc Natl Acad Sci 110:4563–4568
188. Yue B (2014) Biology of the extracellular matrix: an overview. J Glaucoma 23:S20–S23
189. Capila I, Linhardt RJ (2002) Heparin-protein interactions. Angew Chem Int Ed 41:390–412
190. Macri L, Silverstein D, Clark RAF (2007) Growth factor binding to the pericellular matrix and its importance in tissue engineering. Adv Drug Deliv Rev 59:1366–1381
191. Billings PC, Yang E, Mundy C, Pacifici M (2018) Domains with highest heparan sulfate-binding affinity reside at opposite ends in BMP2/4 versus BMP5/6/7: implications for function. J Biol Chem 293:14371–14383
192. Kim T, Yun Y, Park Y, Lee S (2014) In vitro and in vivo evaluation of bone formation using solid freeform fabrication-based bone morphogenic protein-2 releasing PCL/PLGA scaffolds. Biomed Mater 9:025008
193. Martino MM, Tortelli F, Mochizuki M (2011) Engineering the growth factor microenvironment with fibronectin domains to promote wound and bone tissue healing. Sci Transl Med 3:100ra89
194. Sakiyama-elbert SE, Hubbell JA (2000) Development of fibrin derivatives for controlled release of heparin-binding growth factors. J Control Release 65:389–402
195. Jha AK, Mathur A, Svedlund FL (2015) Molecular weight and concentration of heparin in hyaluronic acid-based matrices modulates growth factor retention kinetics and stem cell fate. J Control Release 209:308–316
196. Vieira S, Vial S, Reis RL, Oliveira JM (2017) Nanoparticles for bone tissue engineering. Biotechnol Prog 33:590–611

197. Chiellini F, Piras AM, Errico C (2008) Micro/nanostructured polymeric systems for biomedical and pharmaceutical applications. Nanomedicine 3:367–393
198. Wang Z, Wang K, Lu X (2014) BMP-2 encapsulated polysaccharide nanoparticle modified biphasic calcium phosphate scaffolds for bone tissue regeneration. J Biomed Mater Res Part A 103:1520–1532
199. Kim B, Yang S, Sang C (2018) Incorporation of BMP-2 nanoparticles on the surface of a 3D-printed hydroxyapatite scaffold using an ε -polycaprolactone polymer emulsion coating method for bone tissue engineering. Colloids Surf B 170:421–429
200. Eimori K, Endo N, Uchiyama S, Takahashi Y (2016) Disrupted bone metabolism in long-term bedridden patients. PLoS ONE 11:e0156991
201. Wittkowske C, Reilly GC, Lacroix D, Perrault CM (2016) In vitro bone cell models: impact of fluid shear stress on bone formation. Front Bioeng Biotechnol 4:87
202. Mishra R, Bishop T, Valerio IL (2016) The potential impact of bone tissue engineering in the clinic. Regen Med 11:571–587
203. Volkmer E, Drosse I, Otto S (2008) Hypoxia in static and dynamic 3D culture systems for tissue engineering of bone. Tissue Eng Part A 14:1331–1340
204. Grellier M, Bareille R, Bourget C (2009) Responsiveness of human bone marrow stromal cells to shear stress. J Tissue Eng Regen Med 2:302–309
205. Yourek G, McCormick SM, Mao JJ, Reilly GC (2010) Shear stress induces osteogenic differentiation of human mesenchymal stem cells. Regen Med 5:713–724
206. Singh H, Hutmacher DW (2009) Bioreactor studies and computational fluid dynamics. In: Advances in biochemical engineering/biotechnology series, pp 231–250
207. Rauh J, Ph D, Milan F (2011) Bioreactor systems for bone tissue engineering. Tissue Eng Part B Rev 17:263–280
208. Sladkova M, De Peppo GM (2014) Bioreactor systems for human bone tissue engineering. Processes 2:494–525
209. Martin I, Wendt D, Heberer M (2004) The role of bioreactors in tissue engineering. Trends Biotechnol 22:10–12
210. Gaspar DA, Gomide V, Monteiro FJ (2012) The role of perfusion bioreactors in bone tissue engineering the role of perfusion bioreactors in bone tissue engineering. Biomatter 2:1–9
211. Sikavitsas VI, Bancroft GN, Mikos AG (2002) Formation of three-dimensional cell/polymer constructs for bone tissue engineering in a spinner flask and a rotating wall vessel bioreactor. J Biomed Mater Res 62:136–148
212. Meinel L, Karageorgiou V, Fajardo R (2004) Bone tissue engineering using human mesenchymal stem cells: effects of scaffold material and medium flow. Ann Biomed Eng 32:112–122
213. Kim HJ, Kim U, Leisk GG (2007) Bone regeneration on macroporous aqueous-derived silk 3-D scaffolds. Macromol Biosci 7:643–655
214. Stiehler M, Bunger C, Baatrup A (2008) Effect of dynamic 3-D culture on proliferation, distribution, and osteogenic differentiation of human mesenchymal stem cells. J Biomed Mater Res, Part A 89:96–107
215. Song K, Liu T, Cui Z (2007) Three-dimensional fabrication of engineered bone with human bio-derived bone scaffolds in a rotating wall vessel bioreactor. J Biomed Mater Res, Part A 86A:323–332
216. Wang T, Wu H, Wang H (2009) Regulation of adult human mesenchymal stem cells into osteogenic and chondrogenic lineages by different bioreactor systems. J Biomed Mater Res, Part A 88A:935–946
217. Mccoy RJ, Eng D, Brien FJO, Ph D (2010) Influence of shear stress in perfusion bioreactor cultures for the development of three-dimensional bone tissue constructs: a review. Tissue Eng Part B 16:587–601
218. Wang Y, Uemura T, Dong J (2003) Application of perfusion culture system improves in vitro and in vivo osteogenesis of bone marrow-derived osteoblastic cells in porous ceramic materials. Tissue Eng 9:1205–1214
219. Bancroft GN, Sikavitsas VI, Mikos AG (2003) Design of a flow perfusion bioreactor system for bone tissue-engineering applications. Tissue Eng 9:549–554

220. Bhaskar B, Owen R, Bahmaee H (2017) Design and assessment of a dynamic perfusion bioreactor for large bone tissue engineering scaffolds. Appl Biochem Biotechnol 185:555–563
221. Bouet G, Marchat D, Cruel M (2015) In vitro three-dimensional bone tissue models: from cells to controlled and dynamic environment. Tissue Eng Part B 21:133–156
222. Nokhbatolfoghahaei H, Rad MR, Khani M-M (2017) Application of bioreactors to improve functionality of bone tissue engineering constructs: a systematic review. Curr Stem Cell Res Ther 12:564–599
223. Matziolis D, Tuischer J, Matziolis G (2011) Osteogenic predifferentiation of human bone marrow-derived stem cells by short-term mechanical stimulation. Open Orthop J 5:1–6
224. Bölgen N, Yang Y, Korkusuz P (2008) Three-dimensional ingrowth of bone cells within biodegradable cryogel scaffolds in bioreactors at different regimes. Tissue Eng Part A 14:1743–1750
225. Aaron RK, Ciombor DM, Simon BJ (2004) Treatment of nonunions with electric and electromagnetic fields. Clin Orthop Relat Res 02906:21–291
226. Chalidis B, Sachinis N, Hospital SM, Hospital AG (2011) Stimulation of bone formation and fracture healing with pulsed electromagnetic fields: biologic responses and clinical implications. Int J Immunopathol Pharmacol 24:17–20
227. Funk RHW, Monsees T, Nurdan O (2009) Electromagnetic effects-from cell biology to medicine. Prog Histochem Cytochem 43:177–264
228. Sun L, Hsieh D, Lin P (2010) Pulsed electromagnetic fields accelerate proliferation and osteogenic gene expression in human bone marrow mesenchymal stem cells during osteogenic differentiation. Bioelectromagnetics 31:209–219
229. Tsai M, Chang WH, Chang K (2007) Pulsed electromagnetic fields affect osteoblast proliferation and differentiation in bonetissue engineering. Bioelectromagnetics 28:519–528
230. Fassina L, Visai L, De Angelis MGC (2007) Surface modification of a porous polyurethane through a culture of human osteoblasts and an electromagnetic bioreactor. Technol Heal Care 15:33–45
231. Liu C, Abedian R, Meister R (2012) Influence of perfusion and compression on the proliferation and differentiation of bone mesenchymal stromal cells seeded on polyurethane scaffolds. Biomaterials 33:1052–1064
232. Petri M, Ufer K, Toma I (2012) Effects of perfusion and cyclic compression on in vitro tissue engineered meniscus implants. Knee Surg, Sport Traumatol Arthrosc 20:223–231
233. Kang KS, Hong JM, Jeong YH (2014) Combined effect of three types of biophysical stimuli for bone regeneration. Tissue Eng Part A 20:1767–1777
234. Huang R-L, Liu K, Li Q (2016) Bone regeneration following the in vivo bioreactor principle: is in vitro manipulation of exogenous elements still needed? Regen Med 11:475–481
235. Huang R, Kobayashi E, Liu K, Li Q (2016) Bone graft prefabrication following the in vivo bioreactor principle. EBioMedicine 12:43–54
236. Tatara AM, Wong ME, Mikos AG (2014) In vivo bioreactors for mandibular reconstruction. J Dent Res 93:1196–1202
237. Stevens MM, Marini RP, Schaefer D (2005) In vivo engineering of organs: the bone bioreactor. Proc Natl Acad Sci 102:11450–11455
238. Holt GE, Halpern JL, Dovan TT (2005) Evolution of an in vivo bioreactor. J Orthop Res 23:916–923
239. Huang R-L, Tremp M, Ho C-K (2017) Prefabrication of a functional bone graft with a pedicled periosteal flap as an in vivo bioreactor. Sci Rep 7:1–11
240. Zhang H, Mao X, Zhao D (2017) Three dimensional printed polylactic acid-hydroxyapatite composite scaffolds for prefabricating vascularized tissue engineered bone: an in vivo bioreactor model. Sci Rep 7:1–13
241. Akar B, Tatara AM, Sutradhar A (2018) Large animal models of an in vivo bioreactor for engineering. Tissue Eng 24:317–325
242. Cheng M-H, Brey EM, Allori AC (2009) Periosteum-guided prefabrication of vascularized bone of clinical shape and volume. Plast Reconstr Surg 124:787–795

243. Brey EM, Cheng M-H, Allori A (2007) Comparison of guided bone formation from periosteum and muscle fascia. Plast Reconstr Surg 119:1216–1222
244. Cheng M, Brey EM, Ph D (2005) Ovine model for engineering bone segments. Tissue Eng 11:214–225
245. Warnke PH, Springer ING, Wiltfang J (2004) Growth and transplantation of a custom vascularised bone graft in a man. Lancet 364:766–770
246. Wiltfang J, Rohnen M, Egberts A-H (2016) Man as a living bioreactor: prefabrication of a custom vascularized bone graft in the gastrocolic omentum. Tissue Eng Part C Methods 22:740–746
247. Warnke PH, Kosmahl M, Russo PAJ (2006) Man as living bioreactor: fate of an exogenously prepared customized tissue-engineered mandible. Biomaterials 27:3163–3167

# Chapter 14
# Fabrication and In Vitro Corrosion Characterization of 316L Stainless Steel for Medical Application

Kanishka Jha, Jagesvar Verma and Chander Prakash

## 1 Introduction

Today, we are living in an era where defected or damaged musculoskeletal parts are being replaced or repaired by various medical implants manufactured from materials that are biocompatible or biomaterials and show consistency over a longer period of time [1, 2]. Therefore, it is an important aspect of material science which deals with the development of porous solids for prosthetics applications [3]. Apart from the breakthrough earned by various researchers in the area of biomaterials, attaching the implant to the parent bone still causes several issues [4]. Major issue which was cited by many authors is the incongruity in Young's moduli of the implant and the contacting bone, which results in implant slackening following stress shielding behavior of bone [5]. Over a period of time, implanted material has to withstand various physiological forces thus it must have comparable strength and durability. For applicability, biomaterials must adhere to some specifications, like high strength and compatibility with body tissues and fluids. Importance of porous structure is that they are designed in such a manner that they are interconnected and have suitable shape and size to allow attachment and ingrowth of body tissues [6].

Proper selection of the materials for bio-implants is crucial as the material has to meet several requirements for long-term use. Such major requirements are: (i) Material must have biocompatibility, i.e., they must be non-toxic, non-inflammatory for human body; (ii) Their durability is particularly a function of corrosion and wear, and it is also documented that longer contact to human tissue may trigger sensitivity reaction and also lead in generating tumors [7, 8]; and (iii) Acceptable mechanical properties are needed to minimize the fatigue failure after numerous loading of cyclic. Steel serves almost every requirement for biomaterials and also holds the longest record for using as a biomaterial implant [9]. In vast

K. Jha · J. Verma · C. Prakash (✉)
School of Mechanical Engineering, Lovely Professional University, Phagwara, Punjab, India
e-mail: chander.mechengg@gmail.com

© Springer Nature Singapore Pte Ltd. 2019
P. S. Bains et al. (eds.), *Biomaterials in Orthopaedics and Bone Regeneration*,
Materials Horizons: From Nature to Nanomaterials,
https://doi.org/10.1007/978-981-13-9977-0_14

domain of steel, austenitic 316L is the only grade which is used for biocompatibility, and the reason for being alone in family and available for biomaterial application is that it is economic worldwide and non-ferromagnetic in nature [10–14].

## 2    Manufacturing Methods of 316L

### 2.1    Sintering

Sintering one of the widely and simplest techniques used for producing porous metal structures based on partial densification by powder metallurgy. This technique matured over a period of time and is used for both porous coating and fully porous material application as bio-implants. It works with three basic stages, i.e., compacting, binding and sintering metal powder. This technique is a high-temperature process, where the temperature makes the particle to bond with each other and also facilitating no considerable change in shape. To enhance the bonding strength between particles or to increase the compatibility, a binder is often added with a defined particular motive. Figure 1 shows the schematic of fusion of powder particle boundaries during the sintering process [4].

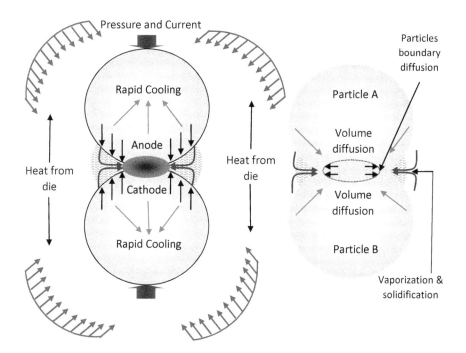

**Fig. 1**  Schematic representation of sintering of powder particles

Sintering of metallic materials by powder metallurgy has efficiently replaced steels produced by conventional methods, like casting or forging, for many several years. This is determined, most importantly, by a huge decrease of production costs, fabricating precision, surface smoothness is widely required to fix, supplant or change diseased or damaged parts of the musculoskeletal framework and should be made from materials that are biocompatible and stable over an extensive stretch of time [15]. The bio-implant alloyed is generally manufactured from Co/Cr-based alloys, stainless steels (SS), titanium and its alloys [16–18]. However, often implant failures take place, in which 42% of failures are because of fracture, 24% was due to corrosion and 14% caused because of the vicious organization along with the implants [19]. 316L SS is considered as a very common material as it is a readily available implant material, economical cost, efficient corrosion and fatigue resistance and ease of fabrication and welding, making this material attractive for surgical bio-implants [20].

## 2.2  Space Holder Method

In comparison to sintering, the space holder method fabricates part with greater porosity. A schematic of the space holder method with various stages is shown in Fig. 2. Entire process is divided into four stages followed by the final product. In Stage 1, raw material in the form of metal powders is taken. Stage 2 depicts the uniform mechanical stirring of metal powder by appropriate space holder material. Mixing was followed by compacting of powder in Stage 3 to form a green body, which was followed by low-temperature heat treatment process specially designed to remove space holder, and also the initiation of sintering of powder was begun with neck formation between two particles. Further sintering at elevated temperatures develops the sinter neck growth. This leads to the densification of the structure and associated improvement of structural integrity. For the structure to maintain its geometry and sufficient mechanical strength, the compaction of metal and space holder material powder must be high enough [21].

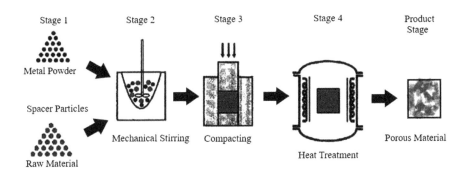

**Fig. 2**  Schematics for space holder method

The basic issue associated with this process is the removal of space holder material from the compacted mixture. Bram et al. [22], for removing this difficulty used urea powder, which could be removed at a temperature below than 200 °C and also reported with very less contamination of the titanium particles. Further sintering was carried at 1400 °C for 1 h to obtain the porosities of the order of 60–70% with a pore diameter in the range of 0.1–0.24 mm, which was a function of urea particle size. In another study, Wen et al. [23] used ammonium hydrogen carbonate as a space holder material for porosity with the decomposition of 200 °C. It is then sintered at 1200 °C for approximately 2 h. The resultant structure contains the porosity of 78% and compressive strength of 35 MPa, and these values are really close to the properties of bone.

## 2.3  Replication

This technique requires a three-stage process where the initial stage corresponds to pattern preparation followed by the intermediate stage of infiltration and then the final stage of removal of pattern. Replication technique is effective because of its flexibility for fabricating various structure designs at considerable low cost [24]. Figure 3 shows the schematic represents the processing route of replication technique. Apart from these properties, this technique can achieve high porosity with good channel connection between pores [24, 25], which supports the bone in growth and vascularization of budding tissues [26]. High porosity and interconnection find applications in various areas like dental implants, hip prostheses, permanent osteosynthesis, etc. [27]. Li et al. [29] prepared the samples with sponge replication technique as shown in figure in the three-stage process. Samples were initially heated at very low rates (2 °C/min) to 400 °C for 2 h. The open porosity was found to be more than 74% which points maximum connected pores. Porous titanium and its alloy structures were also fabricated by Li et al. with this method [28]. They immersed polyurethane foam in

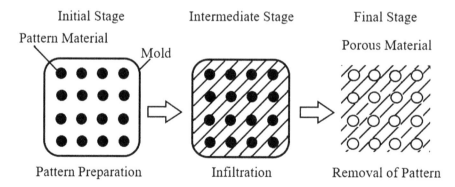

**Fig. 3** Schematics representation of replication

titanium slurry (Ti–6Al–4V and $H_2O$) and ammonia solution. All specimens were dried, and the procedure was repeated until entire polyurethane foam was coated with titanium powder. Further, the specimens were heated and ensured the removal of polyurethane scaffolds, and after subsequent sintering, porous titanium alloy was fabricated with 88% porosity and compressive strength of 10 MPa.

## 2.4  Combustion Synthesis

One of the most recent and effective techniques for fabricating high purity porous alloy is the combustion synthesis. Combustion drives the particle's fusion by releasing a large amount of heat in a self-sustaining exothermic reaction. Heating process through exothermic reaction is done in two stages: Initially, the samples heated gradually to allow throughout heating in Stage 2. Self-sustaining elevated temperature synthesis proceeds with a combustion wave propagating through the entire mixture. Samples porosity and microstructure were influenced by various processing parameters like particle size, binder and compaction pressure. Fabrication of nitinol was also conducted by numerous researchers using combustion synthesis. Li et al. made porous NiTi shape memory alloys having anisotropic pore structure by SHS [29]. Titanium having 21 mm and nickel having 38 mm size powders were mixed together and cold pressed which further placed in a reaction chamber which was ignited at a preheating temperature of 550 °C by using heating element (tungsten coil). Figure 4 shows the schematic represents the processing route of combustion synthesis technique.

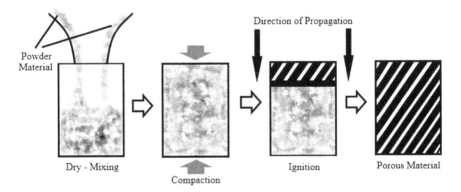

**Fig. 4**  Schematics representation of combustion synthesis

# 3 In Vitro Corrosion Behavior of 316L

In the biomedical field, the gradual degradation of biomaterials is a paramount concern mainly when a metal implant is used in the aggressive electrolytic/chemical environment of the human body. The implants experience harsh corrosion especially in human blood and other parts of the body fluid because in the body fluid system huge elements are present such as sodium, water, chlorine, amino acids, saliva and proteins [30]. In the human body, the aqueous medium comprises a variety of anions like chloride, bicarbonate ions, phosphate and cations such as $Mg^{2+}$, $K^+$, $Na^+$, $Ca^{2+}$, etc. [30, 31]. These above-mentioned constituents upset the equilibrium states by anodic and/or cathodic reaction in the metallic implant and consume it. Proteins can attach with metal ions and take it away from the metallic implant surface and disturbing the equilibrium.

Many researchers have reported that the solid 316L and/or 316L porous austenitic stainless steels (ASSs) are extensively used for orthopedic applications owing to its combination of better mechanical properties [32]. Nevertheless, it is more prone to corrosion in body fluids [30, 31]. In general, the 316L ASS corrosion resistance is mainly due to the formation of passive film. It is distinguished that the stainless steels made by sintering process mainly have low resistance of resistance than stainless steels developed by the conventional metallurgical process, due to the intrinsic porosity of the first [30]. However, these porosities amplify the crevice and/or pitting corrosion [31]. Recently, the movement is toward the focusing attention in the corrosion behavior of porous 316L ASSs with various alloying constituent and porosity [33]. However, in the porosities as corrosion concerned, crevice corrosion may take place [4] because passivity of the alloy reduced by concentration cells' formation within the pores [33].

Failure or removed stainless steel implants from the human body often shows sign of crevice or pitting corrosion. Fatigue and corrosion combined effect may also lead to the failure of metal implants [34]. Moreover, several authors reported the issues and influence of corrosion products which leached from neighboring tissues in the surrounding area of the implants. This interaction mainly caused infectious and/or allergic reactions in the human body which may also case to the failure of the biomedical device [34, 35]. Many researchers are using surface modifications on the biomedical implants to enhance the biocompatibility, corrosion and wear resistance [34–37].

This literature incorporates the in vitro corrosion behavior of solid 316L and porous. The corrosion behavior of 316L ASS mainly depends upon the passive layer of chromium oxide. However, numeral research has established that the irregular and porous surface geometry of porous 316L ASSs showed higher corrosion rates in vitro [34–36]. Seah et al. [38] reported that the corrosion resistance decreased with decreasing porosity. Authors concluded that the porous morphology traps the ionic species and hampers the entrée of oxygen, because oxygen mainly causes the development of corrosion resistant layers. In in vitro test, the corrosion performance of the implants is simulated with the human body fluid environment such as

Hank's and Ringer's. However, the test is performed to simulate the human body in controlled temperature (~37 °C) and pH values approximate 7.35–7.45 [38]. Genotoxicity tests should also perform as per the standard such as ISO 10993 (part 3), the sample should be prepared to make contact with human tissues, whether in a transitory or permanent use [39]. Abdullah [40] studied the potentiodynamic polarization test on the porous 316L ASSs to study the corrosion behavior. Anodic polarization (Ecorr) curve is recorded in −0.40 to −0.60 v and breakdown potential (Eb) is −0.3 to −0.4 v in NaCl solution. It was reported from the SEM micrograph that mainly three types of formation occurred on to the surface due to corrosion reaction includes corrosion pits, the formation of branching and clustered crystallite structure and the formation of the particles of corrosion product. It was concluded that the pore structure and morphology of the porous 316L ASS influence the corrosion attack. Mariotto et al. [41] have performed the electrochemical experiments using a NaCl corrosive medium, 316L ASS foams and it is compared with 316L massive steel. Authors reported that the 316L massive steel had good corrosion behavior compared to foams and also concluded that the porosity was the main factors to increase the surface area exposed to the corrosive NaCl aqueous solution. Costa et al. have tested and reported that the 316L ASS, which was prepared by either powder injection molding (PIM) technique or conventional root metallurgy, have high resistance for corrosion in the Eagle's medium (MEM) [42]. Authors also reported that in both tests ASS were not indicating cytotoxic and that it can be considered as good materials for the application of biomedical field in Eagle's medium (MEM).

Gregorutti et al. [43] have studied the corrosion behavior of porous 316L implant taken by the gel-casting process in NaCl solution. It was reported that the examination of the curves showed that dense 316L illustrated a lower rate of corrosion and a broad passive region (near to 450 mV/SCE) than that of porous specimen. The higher rate of corrosion in these specimens, at least by one order of magnitude, possibly will be attributed to the increased electrochemical activity and higher exposed specific area, resulting from the porosity. Qingbo et al. [44] studied the porosity effects on the corrosion performance of sintered 316L ASSs' fiber porous felts with various porosities (60, 70, 80 and 90%) were tested in 5 and 30% $H_2SO_4$ solution and the mass losses were measured. After the immerging, the samples in the solution of sulfuric acid for 24 and 48 h which were cleaned, dried and weighted are shown in Fig. 5. The results were observed that the mass loss increased as the porosity increased. The reason is that the specimen with higher porosity has a higher surface area; hence, the sulfuric acid solution has more contact with the felt surface, resulting in a severer attack of corrosion.

Nagarajan et al. [45] studied the electrochemical behavior of 316L SS which was coated with porous niobium oxide in simulated body fluid (SBF) solution for the orthopedic applications. Solgel method was used to prepare niobium oxides and coated on 316L SS. It was reported from the electrochemical characterization that the coated ASSs showed lower passive current density and nobler corrosion potential compared to uncoated 316L SS in SBF solution. Author suggested this coated ASSs for an alternative solution in orthopedic applications. Čapek [46] studied the properties of a highly porous 316L SS scaffold made by SLM, mainly investigated

(a)                                    (b)

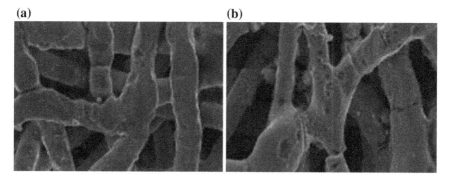

**Fig. 5** Scanning electron microscope of corrosion morphology of sintered SS fiber porous in 30% $H_2SO_4$ at **a** 24 h, and **b** 48 h [44]

surface chemistry and cytocompatibility test and that were compared with non-porous 316L SS. Authors were reported that the observation of cytocompatibility after one day that was similar to wrought 316L SS, which is a commonly used biomaterial. Oshida [47] studied the relationship between the 316L alloying elements' distribution and the tissue response. Authors reported that around Cu and Ni implants severe damage is evident; whereas, around the Fe implant fibrous connective tissue was formed. The dissolved concentration was measured ~10 to 20 Mm. Xu et al. [48] studied the corrosion of 316L SS in simulated body fluids at 37 °C with albumin and $H_2O_2$ in physiological saline to simulate peri-implant condition. Authors reported that the albumin increased the speed of anodic dissolution and repressed cathodic reaction. Presence of albumin decreased pitting potential. However, the addition of $H_2O_2$ increased metal release, mostly Ni release. Brooks [49] assessed the persuade of a simulated (150 mM $H_2O_2$ solution in phosphate-buffered saline) inflammatory response on the 316L ASS corrosion. Specimen deepened in an electrolyte which represented either normal and/or inflammatory physiological conditions. 316L ASS increased corrosion by localized crevice corrosion mechanism when experienced in simulated inflammatory conditions. Author suggested the caution when using 316L ASS.

Duraipandy et al. [50] studied the 316L coated with porous nanostructured Zn incorporated $Nb_2O_5$, (made by sol-gel method) in simulated body fluid (SBF) solution. Authors reported that it improved corrosion resistance, antibacterial activity and suggested/promoted for Osseointegration and biomedical application. Yang [51] studied the addition of Lanthanum (having anticoagulative and antiphlogistic function) into the 316L ASS to enhance its biocompatibility. The corrosion resistance of the Lanthanum added 316L ASS simulated in two different simulated body fluids (1) blood plasma and (2) Hank's solution. Authors reported improved corrosion resistance. Hryniewicz et al. [52] investigated the corrosion behavior of modified 316L ASS surface by magnetoelectropolishing and compared with standard/conventional process conditions. Authors reported that after magnetoelectropolishing process, the sample showed decreased corrosion rate in Ringer's body fluid as compared to the standard/conventional process. Recently, electric discharge machining (EDM) has been found as a potential technique for the surface modification of metallic implants [53–58]. EDM was used to enhance the corrosion resistance, wear resistance, fatigue performance and bioactivity of metallic alloys [59–62]. EDM was also used for the surface modification of SS-316L for the biomedical applications [63–65].

## 4  Conclusion

(1) Corrosion studies and understanding on different body fluid environment are prime importance.
(2) In the biomedical field, porous 316L is more popular, but the corrosion behavior of this material is poor in different body fluid environment.
(3) Different coatings are used on 316L solid metal as well as porous metal to improve corrosion behavior.
(4) Magnetoelectropolishing is also one of the surface modification methods which helps to improve the corrosion resistance.

## References

1. Hollister SJ (2005) Porous scaffold design for tissue engineering. Nat Mater 4(7):518–524
2. Prakash C, Kansal HK, Pabla BS, Puri S, Aggarwal A (2016) Electric discharge machining–a potential choice for surface modification of metallic implants for orthopedic applications: a review. Proc Inst Mech Eng, Part B: J Eng Manuf 230(2):331–353
3. Prakash C, Kansal HK, Pabla BS, Puri S (2017) On the influence of nanoporous layer fabricated by PMEDM on β-Ti implant: biological and computational evaluation of bone-implant interface. Mater Today: Proc 4(2):2298–2307
4. Prakash C, Singh S, Gupta M, Mia M, Królczyk G, Khanna N (2018) Synthesis, characterization, corrosion resistance and in-vitro bioactivity behavior of biodegradable Mg–Zn–Mn–(Si–HA) composite for orthopaedic applications. Materials 11(9):1602
5. Prakash C, Singh S, Pabla BS, Sidhu SS, Uddin MS (2018) Bio-inspired low elastic biodegradable Mg-Zn-Mn-Si-HA alloy fabricated by spark plasma sintering. Mater Manufact Processes, 1–12
6. Prakash C, Singh S, Pabla BS Uddin MS (2018) Synthesis, characterization, corrosion and bioactivity investigation of nano-HA coating deposited on biodegradable Mg-Zn-Mn alloy. Surf Coat Technol 346:9–18. https://doi.org/10.1016/j.surfcoat.2018.04.035
7. Weber JN, White EW, Lebiedzik J (1971) New porous biomaterials by replication of echinoderm skeletal microstructures. Nature 233:337–339
8. Walczak J, Shahgaldi F, Heatley F (1998) In vivo corrosion of 316L stainless-steel hip implants: morphology and elemental compositions of corrosion products. Biomaterials 19(1–3):229–237
9. Case C, Langkamer V, James C et al (1994) widespread dissemination of metal debris from implants. Bone Joint J 76(5):701–712
10. Kang C-W, Fang F-Z (2018) State of the art of bioimplants manufacturing: part I. Adv Manuf 6:20–40
11. Lemons J, Niemann K, Weiss A (1976) Biocompatibility studies on surgical-grade titanium-, cobalt-, and iron-base alloys. J Biomed Mater Res, Part A 10(4):549–553

12. Escalas F, Galante J, Rostoker W et al (1976) Biocompatibility of materials for total joint replacement. J Biomed Mater Res, Part A 10(2):175–195
13. Syrett BC, Davis EE (1979) In vivo evaluation of a high strength, high-ductility stainless steel for use in surgical implants. J Biomed Mater Res, Part A 13(4):543–556
14. Breme H, Biehl V, Helsen J (1998) Metals and implants. Metals Biomater 615(46):37–72
15. David Y (1999) The biomedical engineering handbook. In: Bronzino JD (ed) The biomedical engineering handbook, 2nd edn. 2 Volume Set. CRC Press, Boca Raton
16. Chena F-M, Liuc X (2016) Advancing biomaterials of human origin for tissue engineering. Prog Polym Sci 53:86–168
17. Bhui AS, Singh G, Sidhu SS, Bains PS (2018) Experimental investigation of optimal ED machining parameters for Ti–6Al–4V biomaterial. FU Mech Eng 16(3):337–345
18. Bains PS, Mahajan R, Sidhu SS, Kaur S (2019) Experimental investigation of abrasive assisted hybrid EDM of Ti–6Al–4V. J Micromanuf. https://doi.org/10.1177/2516598419833498
19. Zhang LC, Attar H (2016) Selective laser melting of titanium alloys and titanium matrix composites for biomedical applications: a review. Adv Eng Mater 18(4):463–475
20. Gaviria L, Salcido JP, Guda T, Ong JL (2014) Current trends in dental implants. J Korean Assoc Oral Maxillofac Surgeons 40(2):50–60
21. Srivatsan TS (2009) Processing and fabrication of advanced materials, XVII: Volume One (vol. 1). IK International Pvt Ltd
22. Ryan G, Pandit A (2006) Dimitrios Panagiotis Apatsidis. Fabrication methods of porous metals for use in orthopaedic applications. Biomaterials 27:2651–2670
23. Bram M (2000) High-porosity titanium, stainless steel, and superalloy parts. Adv Eng Mater 2:196–199
24. Wen CE, Yamanda Y, Shimojima K, Chino Y, Asahina T, Mabuchi M (2001) Fabrication and characterization of autogenous titanium foams. Eur Cells Mater 1:61–2
25. Cachinho SCP, Correia RN (2007) Titanium porous scaffolds from precursor powders: rheological optimization of TiH$_2$ slurries. Powder Technol 178:109–113
26. Lee JH, Kim HE, Shin KH, Koh YH (2010) Improving the strength and biocompatibility of porous titanium scaffolds by creating elongated pores coated with a bioactive, nanoporous TiO$_2$ layer. Mater Lett 64:2526–2529
27. Karageorgiou V, Kaplan D (2005) Porosity of 3D biomaterial scaffolds and osteogenesis. Biomaterials 26:5474–5491
28. Barrabés M, Sevilla P, Planell JA, Gil FJ (2008) Mechanical properties of nickel-titanium foams for reconstructive orthopaedics. Mater Sci Eng, A 28:23–27
29. Li JP, Li SH, Groot K, Layrolle P (2002) Preparation and characterization of porous titanium. Key Eng Mater 218–220:51–54
30. Li BY, Rong LJ, Li YY, Gjunter VE (2000) A recent development in producing porous NiTi shape memory alloys. Intermetallics 8:881–884
31. Sharon A, Melman N, Itzhak D (1994) Corrosion resistance of sintered stainless steel containing nickel based additives. Powder Metall 37(1):67–71
32. Mathiesen T, Maalm E (1995) Adv Powder Metall Particul Mater 3:45
33. Peled P, Itzhak D (1990) The corrosion behavior of double pressed, double sintered stainless steel containing noble alloying elements. Corros Sci 30(1):59–65
34. Seah KHW, Thampuran R, Chen X, Teoh SH, Off C (1995) Sci 37(9):1333
35. Ryan G, Pandit A, Apatsidis DP (2006) Fabrication methods of porous metals for use in orthopaedic applications. Biomaterials. 27:2651–2670. PM id:16423390. https://doi.org/10.1016/j.biomaterials.2005.12.002
36. Ducheyne P (1983) In vitro corrosion study of porous metal fibre coatings for bone ingrowth. Biomaterials 4:185–191. https://doi.org/10.1016/0142-9612(83)90008-X8
37. Bains PS, Sidhu SS, Payal HS (2019) Magnetic field influence on surface modifications in powder mixed EDM. Silicon 11(1):415–423
38. Seah KHW, Thampuran R, Teoh SH (1998) The influence of pore morphology on corrosion. Corros Sci 40:547–556. https://doi.org/10.1016/S0010-938X(97)00152-2

39. Asri RIM, Harun WSW, Samykano M, Lah NAC, Ghani SAC, Tarlochan F, Raza MR (2017) Corrosion and surface modification on biocompatible metals: a review. Mater Sci Eng, C 77:1261–1274
40. Abdullah Z, Ismail A, Ahmad S (2017) The influence of porosity on corrosion attack of Austenitic stainless steel. J Phys: Conf Ser 914(1):012013. IOP Publishing
41. Mariotto SDFF, Guido V, Yao Cho L, Soares CP, Cardoso KR (2011) Porous stainless steel for biomedical applications. Mater Res 14(2):146–154
42. Costa I, Rogero SO, Correa OV, Kunioshi CT, Saiki M (2005) Corrosion and cytotoxicity evaluation of AISI 316L stainless steel produced by powder injection molding (PIM) technology. In: Materials science forum. vol 498. Trans Tech Publications, pp 86–92
43. Gregorutti RW, Elsner CI, Garrido L, Ozols A (2015) Corrosion in 316L porous prostheses obtained by gelcasting. Procedia Mater Sci 9:279–284)
44. Qingbo A, Huiping T, Jianzhong W, Hao Z, Jun M, Bin L (2014) Corrosion behavior of sintered 316L stainless steel fibre porous felt. Rare Metal Mater Eng 43(10):2344–2348
45. Nagarajan S, Raman V, Rajendran N (2010) Synthesis and electrochemical characterization of porous niobium oxide coated 316L SS for orthopedic applications. Mater Chem Phys 119(3):363–366
46. Čapek J, Machova M, Fousova M, Kubásek J, Vojtěch D, Fojt J, Ruml T (2016) Highly porous, low elastic modulus 316L stainless steel scaffold prepared by selective laser melting. Mater Sci Eng, C 69:631–639
47. Oshida Y (2007) Implant-related biological reactions. Bioscience and Bioengineering of Titanium Materials, Elsevier, pp 157–214
48. Xu W, Yu F, Yang L, Zhang B, Hou B, Li Y (2018) Accelerated corrosion of 316L stainless steel in simulated body fluids in the presence of H2O2 and albumin. Mater Sci Eng: C
49. Brooks EK, Brooks RP, Ehrensberger MT (2017) Effects of simulated inflammation on the corrosion of 316L stainless steel. Mater Sci Eng, C 71:200–205
50. Duraipandy N, Syamala KM, Rajendran N (2018) Antibacterial effects, biocompatibility and electrochemical behavior of zinc incorporated niobium oxide coating on 316L SS for biomedical applications. Appl Surf Sci 427:1166–1181)
51. Yang H, Yang K, Zhang B (2007) Pitting corrosion resistance of La added 316L stainless steel in simulated body fluids. Mater Lett 61(4–5):1154–1157
52. Hryniewicz T, K Rokosz, Filippi M. Biomaterial Studies on AISI 316L Stainless Steel after Magneto-electropolishing
53. Singh H, Singh S, Prakash C (2019) Current trends in biomaterials and bio-manufacturing. In: Prakash C et al (eds) Biomanufacturing, Springer, Cham, pp 1–34
54. Prakash C, Uddin MS (2017) Surface modification of β-phase Ti implant by hydroaxyapatite mixed electric discharge machining to enhance the corrosion resistance and in-vitro bioactivity. Surf Coat Technol 326 Pt A:134–145
55. Prakash C, Singh S, Pabla BS, Uddin MS (2018) Synthesis, characterization, corrosion and bioactivity investigation of nano-HA coating deposited on biodegradable Mg-Zn-Mn alloy. Surf Coat Technol 346:9–18
56. Prakash C, Singh S, Singh M, Verma K, Chaudhary B, Singh S (2018) Multi-objective particle swarm optimization of EDM parameters to deposit HA-coating on biodegradable Mg-alloy. Vacuum 158:180–190
57. Prakash C, Kansal HK, Pabla BS, Puri S (2015) Potential of powder mixed electric discharge machining to enhance the wear and tribological performance of β-Ti implant for orthopedic applications. J Nanoeng Nanomanuf 5:261–269. https://doi.org/10.1166/jnan.2015.1245
58. Prakash C, Kansal HK, Pabla BS, Puri S (2017) Experimental investigations in powder mixed electrical discharge machining of Ti-35Nb-7Ta-5Zr β-Ti alloy. Mater Manuf Process 32:274–285. https://doi.org/10.1080/10426914.2016.1198018
59. Prakash C, Kansal HK, Pabla BS, Puri S (2015) Processing and characterization of novel biomimetic nanoporous bioceramic surface on β-Ti implant by powder mixed electric discharge machining. J Mater Eng Perform 24:3622–3633. https://doi.org/10.1007/s11665-015-1619-6

60. Prakash C, Kansal HK, Pabla BS, Puri S (2016) Multi-objective optimization of powder mixed electric discharge machining parameters for fabrication of biocompatible layer on β-Ti alloy using NSGA-II coupled with Taguchi based response surface methodology. J Mech Sci Technol 30:4195–4204. https://doi.org/10.1007/s12206-016-0831-0

61. Prakash C, Kansal HK, Pabla BS, Puri S (2015) Powder mixed electric discharge machining an innovative surface modification technique to enhance fatigue performance and bioactivity of β-Ti implant for orthopaedics application. J Comput Inf Sci Eng 14:041006. https://doi.org/10.1115/1.4033901

62. Prakash C, Kansal HK, Pabla BS, Puri S (2016) Effect of surface nano-porosities fabricated by powder mixed electric discharge machining on bone-implant interface: an experimental and finite element study. Nanosci Nanotechnol Lett 8:815–826. https://doi.org/10.1166/nnl.2016.2255

63. Raju P, Sarcar MMM, Satyanarayana B (2014) Optimization of wire electric discharge machining parameters for surface roughness on 316L stainless steel using full factorial experimental design. Procedia Mater Sci 5:1670–1676

64. Nahak Binayaka, Gupta Ankur (2019) A review on optimization of machining performances and recent developments in electro discharge machining. Manufact Rev 6:2

65. Takale AM, Chougule NK (2018) Evaluation of wire electro-discharge machining performance characteristics of Ti49. 4Ni50. 6 shape memory alloy and SS316L for orthopedic implant application. Trends Biomater Artif Organs 32(3):111–117

# Chapter 15
# A Rough Decision-Making Model for Biomaterial Selection

**Dragan Pamucar, Prasenjit Chatterjee, Morteza Yazdani and Shankar Chakraborty**

## 1  Introduction

Biomaterials can be interpreted as the category of materials used in medical devices and their degree of sophistication has increased significantly. The benefits of engineered materials incorporate unsurprising mechanical properties and simplicity of their treatment. Wide range of scientific projects in material engineering files is shifted to biomaterials application. Those materials interact and meet with biological systems requirement for specific medical purposes are categorized as biomaterials. This section of engineering body is extracted from synthetic polymers by a variety of chemical processes utilizing metallic components, polymers, ceramics, or composite materials. It is essential to mention biomaterials are very typical in today's dental applications, surgery, and drug delivery [1, 2]. Bioengineering consists of biological, chemical, tissue engineering medicine, and material science subjects. Range of applications and utilizations like building artificial organs, rehabilitation devices, or implants to replace natural body tissues are encountered in this area. The world of

D. Pamucar
Department of Logistics, Military Academy, University of Defence in Belgrade, Pavla Jurisica Sturma 33, Belgrade, Serbia
e-mail: dpamucar@gmail.com

P. Chatterjee (✉)
Department of Mechanical Engineering, MCKV Institute of Engineering, Howrah, West Bengal, India
e-mail: prasenjit2007@gmail.com

M. Yazdani
Department of Management, Universidad Loyola Andalucía, Andalucía, Spain
e-mail: morteza_yazdani21@yahoo.com

S. Chakraborty
Department of Production Engineering, Jadavpur University, Kolkata, West Bengal, India
e-mail: s_chakraborty00@yahoo.co.in

© Springer Nature Singapore Pte Ltd. 2019
P. S. Bains et al. (eds.), *Biomaterials in Orthopaedics and Bone Regeneration*,
Materials Horizons: From Nature to Nanomaterials,
https://doi.org/10.1007/978-981-13-9977-0_15

material engineering science confronts to a brilliant concept and nowadays, many companies are trying to invest and support projects for the development of new products. An unstructured model for comparing and analyzing biomedical materials for a specific application may lead to huge failure of the product, repeated processes, considerable loss, impairment of tissue functions and overall increasing of the costs. Biomaterials have changed the demands of customers and medical services and are constantly used in human organs for treating heart diseases, coronary angioplasty, orthopaedics applications, and orthodental structures [3, 4]. The biomaterials are classified into metals, polymers, ceramics, composites, and apatite. In order to deal with the problem of human degenerative diseases, investigators effectively rely on the use of artificial or natural biomaterials for reinstating the functions of affected parts. The specification and properties each engineer seeks in a biomaterial must contain the following item: biomechanical compatibility, biocompatibility, high corrosion and wear resistance, and osseointegration [5, 6]. Biomaterials also play a vital role in fabrication of biological screening devices as well as in a large range of non-biomedical applications. Discussions on some common and familiar biomaterials are presented here. One of the vastly used biomaterials is the metallic implants which are the primary materials used for joint replacement and orthopedic applications. Exceptional thermal conductivity, excellent strength, higher fracture toughness, corrosion resistance, and hardness along with biocompatibility are the several promising properties that the metallic alloys and materials possess. Stainless steel is employed to fabricate artificial bone and becomes the predominant implant alloy due to ease in fabrication and having required mechanical properties and corrosion resistance. Cobalt-based alloys, Ti alloys, ceramic materials, zirconia ceramics, and polymeric materials are other well-known types of biomaterials [7–9]. Biomaterials are selected to simultaneously satisfy a broad range of fundamental requirements from mechanical to biological aspects. The construction of the femoral implant must meet several criteria like adequate strength, ductility, elastic modulus, wear resistance, corrosion resistance, biocompatibility, and osseointegration. For example, according to Hafezalkotob and Hafezalkotob [10], density and elastic modulus properties are also examined for compatible designs which certainly are strategically necessary implants and prostheses for material applications. Biocompatibility basically signifies the potentiality and fitness of a material for not being malignant or physiologically sensitive with living organisms. This basic requirement is considered to be the most important issue in the design and selection of implants and the material to be used for its fabrication [11]. The very required process of engineering and design decision making is how to deal with the complex system, decision-making rules, many variables, and parameters. The recognition of this approach not even build a robust decision system, it effectively carries out a quality of results and further approval.

A decision-making system comes up with situations where sort of alternatives (choices) are evaluated with respect to certain factors or criteria. It is valuable to resolve the decision problem with a well-established mechanism to reveal the optimal solution. Although, this mechanism must satisfy the policy maker's viewpoints, however, all conflicting objectives with different optimization direction cause errors and shape uncertain and incorrect conditions. This point leads academic and industrial partners to tolerate pressures which needs extraordinary endeavor. Therefore, it is argued that the key item in almost all of the engineering decisions is to draw the objectives and map the alternative options in order to overcome existing complexities. Certainly, application of conventional methods has been saturated while many engineering sectors are reforming their evaluation and measurement systems. Undoubtedly, choosing advanced methods is highly appreciated and today's material investigators in real projects understand various concepts and logics to adopt a comprehensive formula in order to take more efficient decisions. All in all, the realization of multi-criteria decision-making (MCDM) methods in such kind of situations can refine the question of what methodology fits to what decision problem and in what way. It eliminates the complexity of decision problem in a productive manner and formulates a platform to the assessment and selection of the optimal solution [12–14]. Some MCDM techniques are able to configure the decision problem containing alternatives, factors, and decision makers' (DMs) opinions, break it to hierarchical format, analyze, normalize, and finally find the solution. To name them, analytical hierarchy process (AHP) [15], technique of order preference by similarity to ideal solution (TOPSIS) [16, 17], Vlse Kriterijumska Optimizacija I Kompromisno Resenje (VIKOR) [18–21], complex proportional assessment (COPRAS) [22–26], multi-objective optimization on the basis of ratio analysis (MOORA) [27], evaluation based on distance from average solution (EDAS) [28, 29] and combinative distance-based assessment (CODAS) [30, 31] are some of the examples. However, if the selection methodology is carried out randomly or disorganizedly, there will be the risk of overseeing suitable materials and criteria affecting the entire selection process, Hence, the aim of this chapter is to develop a methodology, based on rough AHP and rough CODAS methods, followed by sensitivity analysis and performance comparison to select the most suited biomaterial for a hip joint prosthesis application.

## 2   Literature Review on Biomaterials Selection

Despite there being many MCDM models for general material selection problems, the literature shows very less amount of works to deal with the problems on biomaterials selection. Thus, the aim of this section is to study the past researchers on biomaterial selection and figuring out their weaknesses and enable the DMs to reduce subjectivity and uncertainty to make a clearer support for a strong framework. Bahraminasab and Jahan [32] designed a comprehensive biomaterial selection model for femoral component of total knee replacement (TKR) while employing comprehensive VIKOR method. Jahan and Edwards [33] proposed a weighting technique for dependent and target-based criteria in making optimal decision for biomaterials selection and validated its appropriateness with the extended TOPSIS and comprehensive VIKOR methods. Bahraminasab et al. [34] conducted a multi-objective design optimization process for femoral component of TKR. Petković et al. [35] designed a decision support system while integrating three MCDM tools, i.e., TOPSIS, VIKOR, and weighted aggregated sum product assessment (WASPAS) methods for identifying the best biomaterial alternative for bone implants which could compensate the missing part of a long bone. Hafezalkotob and Hafezalkotob [36] explored the application of comprehensive MULTIMOORA method for hip and knee joint prosthesis materials selection. Chowdary et al. [37] proposed a strategy to prioritize some bioengineering materials under a combined MCDM model with fuzzy AHP and TOPSIS. The research recommends that Polyether ether ketone (PEEK) material is most suitable for biomedical implantations. Kabir and Lizu [38] adopted an integrated FAHP and PROMETHEE methods for selection of femoral material in TKR. Abd et al. [39] employed fuzzy TOPSIS method for hip joint prosthesis material selection. Ristić et al. [40] devised an expert system using fuzzy sets for biomaterial selection in a customized implant application. Hafezalkotob and Hafezalkotob [10] validated the application of interval MULTIMOORA method with target values of attributes based on interval distance and preference degree while utilizing two case studies on hip and knee joint prosthesis materials selection.

## 3 Materials and Methods

### 3.1 Rough Numbers and Operations

Rough numbers (RNs), consisting of the upper, lower, and boundary intervals, determine the intervals of multiple expert evaluations without requiring any additional information and relying only on the original data [41]. Hence, the obtained expert preferences objectively represent and improve the decision-making process. The definition of RNs according to Song et al. [42] is given below.

Let $U$ be a universe containing all the objects and $X$ be a random object from $U$. Then, it is assumed that there exists a set of $k$ classes which represents a DM's preferences, $R = (J_1, J_2, \ldots, J_k)$ with the condition $J_1 < J_2 <, \ldots, < J_k$. Then for every $X \in U, J_q \in R, 1 \leq q \leq k$, the lower approximation $\underline{\text{Apr}}(J_q)$, the upper approximation $\overline{\text{Apr}}(J_q)$, and the boundary interval $\text{Bnd}(J_q)$ are determined as follows:

$$\underline{\text{Apr}}(J_q) = \cup \{X \in U/R(X) \leq J_q\} \tag{1}$$

$$\overline{\text{Apr}}(J_q) = \cup \{X \in U/R(X) \geq J_q\} \tag{2}$$

$$\text{Bnd}(J_q) = \cup \{X \in U/R(X) \neq J_q\}$$
$$= \{X \in U/R(X) > J_q\} \cup \{X \in U/R(X) < J_q\} \tag{3}$$

The object can be represented by a rough number with the lower limit $\underline{\text{Lim}}(J_q)$ and the upper limit $\overline{\text{Lim}}(J_q)$ in Eqs. (4)–(5).

$$\underline{\mathrm{Lim}}(J_q) = \frac{1}{M_{\mathrm{L}}} \sum R(X) | X \in \underline{\mathrm{Apr}}(J_q) \tag{4}$$

$$\overline{\mathrm{Lim}}(J_q) = \frac{1}{M_{\mathrm{U}}} \sum R(X) | X \in \overline{\mathrm{Apr}}(J_q) \tag{5}$$

where $M_{\mathrm{L}}$ and $M_{\mathrm{U}}$ represent the sum of objects given in the lower and upper object approximations of $J_q$, respectively. For object $J_q$, the rough boundary interval ($\mathrm{IRBnd}(J_q)$) is the interval between the lower and upper limits [43]. The rough boundary interval presents a measure of uncertainty. A bigger $\mathrm{IRBnd}(J_q)$ value shows that the variations in experts' preferences exist, while smaller values show that experts' opinions do not differ considerably. All the objects between the lower limit $\underline{\mathrm{Lim}}(J_q)$ and the upper limit $\overline{\mathrm{Lim}}(J_q)$ of the rough number $\mathrm{RN}(J_q)$ are included in $\mathrm{IRBnd}(J_q)$. Since RNs belong to a group of interval numbers, arithmetic operations applied to interval numbers are also appropriate for RNs.

## 3.2   R-AHP Method

As one of the most popular methods of MCDM, the AHP has widely been used for criteria weight estimation in a wide range of applications [43]. AHP is a structured technique for organizing and analyzing complex decisions. It uses the definitions of relative importance to evaluate the weights of the selection criteria. It also bestows flexibility in quantifying the consistency in DMs' preferences in a group decision-making system (GDMS). Due to the presence of uncertainty, subjectivity, and unreliability in GDM, this chapter uses a RN-based AHP method to exploit judgments and imprecision. The succeeding section provides a detail description of the adopted methodology for applying the RN-based AHP method for estimation of criteria weights.

*Step 1. Pairwise comparisons of the criteria:*

Assuming that there exists a group of $m$ experts $\{e_1, e_2, \ldots, e_m\}$ and $n$ criteria $\{c_1, c_2, \ldots c_n\}$, each expert should determine the degree of mutual influence of criteria $i$ and $j (\forall i, j \in n)$. For this purpose, a pairwise comparative analysis of the $i$th and $j$th criteria sets for $k$th expert ($1 \leq k \leq m$) is made and denoted by the values $\xi_{ij}^k (i, j = 1, 2, \ldots, n; k = 1, 2, \ldots, m)$, which were performed using Saaty's 9-point scale [44] and shown by the following matrix.

$$N^{(k)} = \left[ \xi_{ij}^{(k)} \right]_{n \times n} = \begin{bmatrix} \xi_{11}^{(k)} & \xi_{12}^{(k)} & \cdots & \xi_{1n}^{(k)} \\ \xi_{21}^{(k)} & \xi_{22}^{(k)} & \cdots & \xi_{2n}^{(k)} \\ \vdots & \vdots & \ddots & \vdots \\ \xi_{n1}^{(k)} & \xi_{n2}^{(k)} & \cdots & \xi_{nn}^{(k)} \end{bmatrix} ; \quad 1 \leq i, j \leq n; \quad 1 \leq k \leq m \tag{6}$$

where $\xi_{ij}^{(k)}$ are linguistic Equations of Saaty's 9-point scale and $\xi_{ij}^{(k)} = 1$ if $i = j$.

*Step 2. Estimation of weights of the Experts':*

For each matrix $N^{(k)}$, consistency of the expert judgement is verified in two steps using the consistency ratio (CR) concept [45]. At first, consistency index (CI) is computed, followed by which CR is estimated using the standard relationship between CI and the random index (RI). A CR value of less than or equal to 0.10 indicates consistent judgments made by the experts [46].

The weight importance of the experts are now determined using Eqs. (7) and (8).

$$\delta_k = \frac{1}{CR_k}; \quad 1 \le e \le k \tag{7}$$

where $CR_k$ is the CR of the $k$th expert, and $\delta_k$ is the weight of expert $k$ $(1 \le k \le m)$.

$$w_k = \frac{\delta_k}{\sum_{k=1}^{m} \delta_k} \tag{8}$$

where $\delta_k$ is the weight coefficient of the expert $k(1 \le k \le m)$and $w_k$ is normalized weight coefficient of the expert $k$ and $\sum_{k=1}^{m} w_k = 1$.

*Step 3. Construction of an averaged rough comparison matrix (ARCM):*

Using Eqs. (1)–(5), elements $\xi_{ij}^{(k)}$ of comparison matrix $N^{(k)}$ are now transformed into RNs as $RN(\xi_{ij}^{(k)}) = \left[ \underline{Lim}(\xi_{ij}^{(k)}), \overline{Lim}(\xi_{ij}^{(k)}) \right] = \left[ \xi_{ij}^{(k)-}, \xi_{ij}^{(k)+} \right]$, where $\underline{Lim}(\xi_{ij}^{(k)})$ is the lower approximation of the object class $\xi_{ij}^{(k)}$, and $\overline{Lim}(\xi_{ij}^{(k)})$ is the upper approximation. In this way, for each pairwise comparison matrix, rough sequences are obtained as $RN(\xi_{ij}^{(k)}) = \left\{ \left[ \underline{Lim}(\xi_{ij}^{(1)}), \overline{Lim}(\xi_{ij}^{(1)}) \right], \left[ \underline{Lim}(\xi_{ij}^{(2)}), \overline{Lim}(\xi_{ij}^{(2)}) \right], \dots, \left[ \underline{Lim}(\xi_{ij}^{(m)}), \overline{Lim}(\xi_{ij}^{(m)}) \right] \right\}$.

For each matrix $N^{(k)}$, we get rough sequence $RN(\xi_{ij}^{(k)}) = \left[ \underline{Lim}(\xi_{ij}^{(k)}), \overline{Lim}(\xi_{ij}^{(k)}) \right]$ on the position $(i, j)$ and finally by applying Eq. (9), we get the averaged rough number $RN(\xi_{ij}) = \left[ \underline{Lim}(\xi_{ij}), \overline{Lim}(\xi_{ij}) \right] = \left[ \xi_{ij}^{-}, \xi_{ij}^{+} \right]$

$$RN(\xi_{ij}) = RN\left\{ \left[ \xi_{ij}^{(1)-}, \xi_{ij}^{(1)+} \right], \left[ \xi_{ij}^{(2)-}, \xi_{ij}^{(2)+} \right], \dots, \left[ \xi_{ij}^{(m)-}, \xi_{ij}^{(m)+} \right] \right\}$$
$$= \begin{cases} \xi_{ij}^{-} = \prod_{k=1}^{m} \left( \xi_{ij}^{(k)-} \right)^{w_k} \\ \xi_{ij}^{+} = \prod_{k=1}^{m} \left( \xi_{ij}^{(k)+} \right)^{w_k} \end{cases} \tag{9}$$

where $\underline{Lim}(\xi_{ij}^{(k)}) = \xi_{ij}^{(k)-}$ and $\overline{Lim}(\xi_{ij}^{(k)}) = \xi_{ij}^{(k)-}$, respectively, represents the lower and upper approximation of the $RN(\xi_{ij}^{(k)})$.

Based on Eqs. (8) and (9), we obtain averaged rough comparison matrix as:

$$N = \begin{bmatrix} 1 & [\xi_{12}^-, \xi_{12}^+] & \cdots & [\xi_{1n}^-, \xi_{1n}^+] \\ [\xi_{21}^-, \xi_{21}^+] & 1 & \cdots & [\xi_{2n}^-, \xi_{2n}^+] \\ \vdots & \vdots & \ddots & \vdots \\ [\xi_{n1}^-, \xi_{n1}^+] & [\xi_{n2}^-, \xi_{n2}^+] & \cdots & 1 \end{bmatrix}_{n \times n} \tag{10}$$

*Step 4. Computation of priority vector:*

Priority vector $(PV)$ is the rough weight $RN(w_j)$ which is determined for each criterion. We obtain the rough weight coefficient $RN(w_j)$ by applying Eqs. (11)–(13). By applying Eq. (11), the following values are estimated.

$$RN(\xi_{ij}') = \sum_{j=1}^{n} RN(\xi_{ij}) = \left[ \sum_{j=1}^{n} \xi_{ij}^-, \sum_{j=1}^{n} \xi_{ij}^+ \right] \tag{11}$$

and dividing matrix elements of $N$ with the values obtained from Eq. (11), the normalized matrix $(W)$ is calculated.

$$RN(\varpi_{ij}) = \left[ \varpi_{ij}^-, \varpi_{ij}^+ \right] = \frac{RN(\xi_{ij})}{\sum_{j=1}^{n} RN(\xi_{ij})} = \frac{\left[ \xi_{ij}^-, \xi_{ij}^+ \right]}{\left[ \sum_{j=1}^{n} \xi_{ij}^-, \sum_{j=1}^{n} \xi_{ij}^+ \right]} \tag{12}$$

A normalized matrix for the rough weights is obtained as follows:

$$W = \begin{bmatrix} 1 & [\varpi_{12}^-, \varpi_{12}^+] & \cdots & [\varpi_{1n}^-, \varpi_{1n}^+] \\ [\varpi_{21}^-, \varpi_{21}^+] & 1 & \cdots & [\varpi_{2n}^-, \varpi_{2n}^+] \\ \vdots & \vdots & \ddots & \vdots \\ [\varpi_{n1}^-, \varpi_{n1}^+] & [\varpi_{n2}^-, \varpi_{n2}^+] & \cdots & 1 \end{bmatrix}_{n \times n} \tag{13}$$

where $RN(\varpi_{ij}) = \left[ \varpi_{ij}^-, \varpi_{ij}^+ \right]$ represents a normalized weights coefficients of matrix (10).

The final rough criteria weights are determined using Eq. (14).

$$RN(w_j) = \left[ \frac{1}{n} \sum_{i=1}^{n} \varpi_{ij}^-, \frac{1}{n} \sum_{i=1}^{n} \varpi_{ij}^+ \right] \tag{14}$$

The criteria weights are placed in the interval $RN(w_j) = \left[ w_j^-, w_j^+ \right]$ where the condition is satisfied that $0 \le w_j^- \le w_j^+ \le 1$ for each evaluation criteria $c_j \in C$ $(C = \{c_1, c_2, \ldots c_n\})$. Since these are rough criteria weights, using Eq. (14), the actual weight coefficients are obtained, where $\sum_{j=1}^{n} w_j^- \le 1$ and $\sum_{j=1}^{n} w_j^+ \ge 1$. It satisfies the conditions $w_j \in [0, 1]$ and $j = 1, 2, \ldots, n$.

## 3.3  Rough CODAS Method

In this section, an extension of the original CODAS method based on RNs (R-CODAS) has been proposed to deal with the associated uncertainties. CODAS is an efficient method, introduced by Ghorabaee et al. [30]. The procedural steps of the proposed R-CODAS method are now presented below [30, 31]:

*Step 1. Construct the basic rough decision matrix* $(\Phi)$:

First step is evaluating $b$ alternatives to $n$ criteria. Evaluation of the alternatives per each criteria by $k(1 \leq k \leq m)$ expert is denoted as $\eta_{ij}^{(k)}$, where $i = 1,\ldots, b; j = 1,\ldots, n$. The judgment of $k$ expert is presented as matrix $\Phi^{(k)} = [\eta_{ij}^{(k)}]_{b \times n}$, where $1 \leq k \leq m$.

$$\Phi^{(k)} = \begin{bmatrix} \eta_{11}^{(k)} & \eta_{12}^{(k)} & \cdots & \eta_{1n}^{(k)} \\ \eta_{21}^{(k)} & \eta_{22}^{(k)} & \cdots & \eta_{2n}^{(k)} \\ \vdots & \vdots & \ddots & \vdots \\ \eta_{b1}^{(k)} & \eta_{b2}^{(k)} & \cdots & \eta_{bn}^{(k)} \end{bmatrix}_{b \times n} ; \quad 1 \leq i \leq b; \quad 1 \leq j \leq n; \quad 1 \leq k \leq m \quad (15)$$

In accordance with this, $\Phi^{(1)}, \Phi^{(2)}, \ldots, \Phi^{(m)}$ matrices are the judgment matrices of each of $m$ experts.

Using Eqs. (1)–(5), each sequence $\eta_{ij}^{(k)}(i = 1, 2, \ldots, b; j = 1, 2, \ldots, n)$ of $\Phi^{(k)} = [\eta_{ij}^{(k)}]_{b \times n}$ is transformed into rough sequence $RN(\eta_{ij}^{(k)}) = \left[\underline{Lim}(\eta_{ij}^{(k)}), \overline{Lim}(\eta_{ij}^{(k)})\right] = \left[\eta_{ij}^{(k)-}, \eta_{ij}^{(k)+}\right]$, where $\underline{Lim}(\eta_{ij}^{(k)}) = \eta_{ij}^{(k)-}$ and $\overline{Lim}(\eta_{ij}^{(k)}) = \eta_{ij}^{(k)+}$ represent lower and upper limits of the rough sequence $RN(\eta_{ij}^{(k)})$, respectively.

For each matrix $\Phi^{(k)} = [\eta_{ij}^{(k)}]_{b \times n}$, we get the rough sequence $RN(\eta_{ij}^{(k)}) = \left[\underline{Lim}(\eta_{ij}^{(k)}), \overline{Lim}(\eta_{ij}^{(k)})\right] = \left[\eta_{ij}^{(k)-}, \eta_{ij}^{(k)+}\right]$ on the position $(i, j)$ and finally by applying Eq. (16), we get the averaged RN $RN(\eta_{ij}) = \left[\underline{Lim}(\eta_{ij}), \overline{Lim}(\eta_{ij})\right] = \left[\eta_{ij}^-, \eta_{ij}^+\right]$

$$RN(\eta_{ij}) = RN\left\{\left[\eta_{ij}^{(1)-}, \eta_{ij}^{(1)+}\right], \left[\eta_{ij}^{(2)-}, \eta_{ij}^{(2)+}\right], \ldots, \left[\eta_{ij}^{(m)-}, \eta_{ij}^{(m)+}\right]\right\}$$
$$= \begin{cases} \eta_{ij}^- = \frac{1}{m} \sum_{k=1}^{m} \eta_{ij}^{(k)-} \\ \eta_{ij}^+ = \frac{1}{m} \sum_{k=1}^{m} \eta_{ij}^{(k)+} \end{cases} \quad (16)$$

where $\underline{Lim}(\eta_{ij}^{(k)}) = \eta_{ij}^{(k)-}$ and $\overline{Lim}(\eta_{ij}^{(k)}) = \eta_{ij}^{(k)-}$, respectively, represent the lower and upper approximation of the $RN(\eta_{ij}^{(k)})$.

Based on Eqs. (3) and (16), we obtain the following ARCM:

$$\Phi = \begin{bmatrix} \left[\eta_{11}^-, \eta_{11}^+\right] & \left[\eta_{12}^-, \eta_{12}^+\right] & \cdots & \left[\eta_{1n}^-, \eta_{1n}^+\right] \\ \left[\eta_{21}^-, \eta_{21}^+\right] & \left[\eta_{22}^-, \eta_{22}^+\right] & \cdots & \left[\eta_{2n}^-, \eta_{2n}^+\right] \\ \vdots & \vdots & \ddots & \vdots \\ \left[\eta_{b1}^-, \eta_{b1}^+\right] & \left[\eta_{b2}^-, \eta_{b2}^+\right] & \cdots & \left[\eta_{bn}^-, \eta_{bn}^+\right] \end{bmatrix}_{b \times n} \tag{17}$$

where $\mathrm{RN}(\eta_{ij}) = \left[\underline{\mathrm{Lim}}(\eta_{ij}), \overline{\mathrm{Lim}}(\eta_{ij})\right] = \left[\eta_{ij}^-, \eta_{ij}^+\right]$ denotes the value of the $i$th alternative for the $j$th criterion ($i = 1, 2, \ldots, b$; $j = 1, 2, \ldots, n$). The matrix elements $\mathrm{RN}(\eta_{ij})$ in Eq. (17) are RNs determined by the experts or by using the aggregation of the experts' decisions.

*Step 2. Normalization of basic matrix element:*

Determine RN normalized decision matrix $N = [\hat{\eta}_{ij}]_{b \times n}$ for beneficial and non-beneficial (cost criteria) criteria by the following expressions:

$$\hat{\eta}_{ij} = \begin{cases} \left[\dfrac{\eta_{ij}^- - \eta_j^-}{\eta_j^+ - \eta_j^-}, \dfrac{\eta_{ij}^+ - x_j^-}{\eta_j^+ - \eta_j^-}\right]; & \text{if } j \in B, \\[2ex] \left[\dfrac{\eta_{ij}^- - \eta_j^+}{\eta_j^- - \eta_j^+}, \dfrac{\eta_{ij}^+ - \eta_j^+}{\eta_j^- - \eta_j^+}\right]; & \text{if } j \in C \end{cases} \tag{18}$$

where $\eta_j^+ = \max_i(\eta_{ij})$, $\eta_j^- = \min_i(\eta_{ij})$, $B$ and $C$ represent the sets of beneficial and non-beneficial criteria, respectively, and $\hat{\eta}_{ij}$ denotes the normalized RN values.

*Step 3. Calculate RN-weighted normalized decision matrix (R):*

The RN-weighted normalized matrix $R = [r_{ij}]_{b \times n}$ is calculated as follows

$$\mathrm{RN}(r_{ij}) = \mathrm{RN}(w_j) \cdot \mathrm{RN}(\hat{\eta}_{ij}) = \left[w_j^-, w_j^+\right] \cdot \left[\hat{\eta}_{ij}^-, \hat{\eta}_{ij}^+\right] \tag{19}$$

where $\mathrm{RN}(w_j) = \left[w_j^-, w_j^+\right]$ represents the final RN criteria weight of $j$th criterion and $\mathrm{RN}(r_{ij}) = \left[r_{ij}^-, r_{ij}^+\right]$ indicates weighted normalized values.

*Step 4. Determine FR negative-ideal solution:*

We obtain the RN negative-ideal solution matrix $\mathrm{NS} = [\mathrm{RN}(ns_j)]_{1 \times n}$ as follows

$$ns_j = \min_i r_{ij} = \left[\min\left\{r_{ij}^-\right\}, \min\left\{r_{ij}^+\right\}\right] \tag{20}$$

*Step 5. Calculate the RN-weighted Euclidean ($ED_i$) and RN-weighted Hamming ($HD_i$) distances of alternatives from the RN negative-ideal solution*

We obtain $ED_i$ and $HD_i$ as per [47, 48] and shown as follows:
RN-weighted Euclidean ($ED_i$) distances:

$$ED_i = \sum_{j=1}^{n} d_E\left(r_{ij}; ns_j\right) \qquad (21)$$

where $d_E\left(r_{ij}; ns_j\right)$ we obtain as follows

$$d_E\left(r_{ij}; ns_j\right) = \sqrt{\dfrac{\left\{r_{ij}^- - ns_j^-\right\}^2 + \left\{r_{ij}^+ - ns_j^+\right\}^2}{2}} \qquad (22)$$

RN-weighted Hamming ($HD_i$) distances

$$HD_i = \sum_{j=1}^{n} d_H\left(r_{ij}; ns_j\right) \qquad (23)$$

where $d_H\left(r_{ij}; ns_j\right)$ we obtain as follows

$$d_H\left(r_{ij}; ns_j\right) = \dfrac{\left|r_{ij}^- - ns_j^-\right| + \left|r_{ij}^+ - ns_j^+\right|}{2} \qquad (24)$$

*Step 6. Determine the relative assessment matrix (RA):*

By applying Eq. (25), we obtain elements of the relative assessment matrix $RA = [p_{ie}]_{b \times b}$

$$p_{ie} = (ED_i - ED_e) + (g(ED_i - ED_e) \times (HD_i - HD_e)) \qquad (25)$$

where $e \in \{1, 2, \ldots, b\}$ and $g$ is a threshold function, defined as follows [49]:

$$g(x) = \begin{cases} 1 \text{ if } |x| \geq \theta \\ 0 \text{ if } |x| < \theta \end{cases} \qquad (26)$$

The threshold parameter ($\theta$) of this function can be set by the DM. In this study, we use $\theta = 0.02$ for the calculations.

*Step 7. Calculate the assessment score ($AS_i$) of each alternative:*

By applying Eq. (27), we obtain assessment score

$$AS_i = \sum_{e=1}^{b} p_{ie} \qquad (27)$$

The alternative with the highest assessment score is the most desirable alternative. Figure 1 illustrates the rough number-based decision-making model in a comprehensive way.

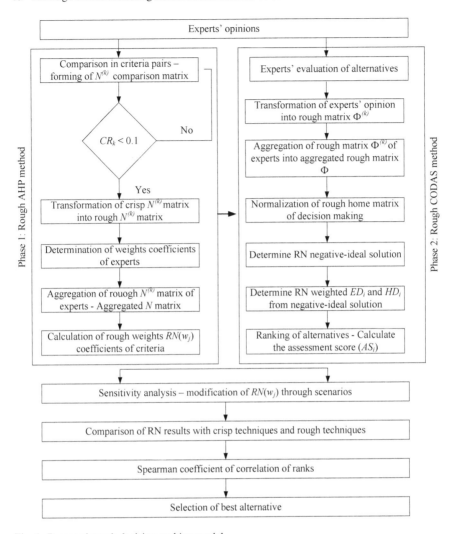

**Fig. 1** Proposed rough decision-making model

## 4 R-AHP-CODAS Model Application

The implementation viability of the proposed R-AHP-CODAS model is now explored via a real-time illustrative example of material selection for femoral element for hip joint prosthesis. Human hip prosthesis has three major parts, namely femoral component, acetabular cup, and acetabular interface. Among these, the femoral element is a rigid metallic rod inserted into the hollow femur. The acetabular cup is normally fitted with ilium. On the other hand, the acetabular interface lies between the femoral component and the acetabular cup. It is made by various materials like metals, polymers, and ceramics to reduce the amount of frictional wear. Evaluation matrix for the considered problem consists of eleven alternative biomaterials and nine criteria which can be found in [39, 50]. Among the nine considered nine attributes, tissue tolerance ($C_1$), corrosion resistance ($C_2$), tensile strength ($C_3$), fatigue strength ($C_4$), toughness ($C_5$), and wear resistance ($C_6$) are the beneficial attributes, whereas elastic modulus ($C_7$), density ($C_8$), and cost ($C_9$) are the attributes regarded as non-beneficial. The concerned study included the expertise of five experts who evaluated the considered alternatives.

### 4.1 Implementation of R-AHP Model

*Step 1. Pairwise comparisons of the criteria:*

After the five experts evaluated each of the nine considered criteria, a comparison matrix between the criteria pairs has been developed, as shown in Table 1.

*Step 2. Determination of weights of the experts:*

After the pairwise comparison, consistency ratio is estimated. Now, using Eqs. (7) and (8), criteria weights are determined, as presented in Table 2.

$$\delta_1 = \frac{1}{CR_1} = \frac{1}{0.0695} = 14.378;$$

$$\delta_2 = \frac{1}{CR_2} = \frac{1}{0.0849} = 11.772;$$

$$\delta_3 = \frac{1}{CR_3} = \frac{1}{0.0980} = 10.208;$$

$$\delta_4 = \frac{1}{CR_4} = \frac{1}{0.0861} = 11.621;$$

$$\delta_5 = \frac{1}{CR_5} = \frac{1}{0.0752} = 13.299.$$

Using Eq. (8), the experts' weights are obtained as follows:

$$w_{E1} = \frac{14.3788}{14.378 + 11.772 + 10.208 + 11.621 + 13.299} = 0.2346;$$

$$w_{E2} = \frac{11.7718}{14.378 + 11.772 + 10.208 + 11.621 + 13.299} = 0.1921;$$

$$w_{E3} = \frac{10.2080}{14.378 + 11.772 + 10.208 + 11.621 + 13.299} = 0.1666;$$

$$w_{E4} = \frac{11.6210}{14.378 + 11.772 + 10.208 + 11.621 + 13.299} = 0.1896;$$

$$w_{E5} = \frac{13.2998}{14.378 + 11.772 + 10.208 + 11.621 + 13.299} = 0.2170.$$

**Table 1** Developed pairwise comparison matrix of the evaluating criteria

| | $C_1$ | $C_2$ | $C_3$ | $C_4$ | $C_5$ | $C_6$ | $C_7$ | $C_8$ | $C_9$ |
|---|---|---|---|---|---|---|---|---|---|
| $C_1$ | 1:1:1:1 | 4:4:7:1:0.1 | 6:9:8:6:5 | 2:2:3:3.3 | 2:5:4:3:2 | 2:5:3:2:2 | 1:3:4:2:3 | 0.3:0.5:0.5:1:0.5 | 1:2:3:2:2 |
| $C_2$ | 0.3:0.3:0.1:1:9 | 1:1:1:1:1 | 4:9:7:4:9 | 3:3:4:3:7 | 3:6:3:6:7 | 3:7:6:4:5 | 2:4:5:7:9 | 0.5:1:0.5:2:3 | 1:4:1:1:4 |
| $C_3$ | 0.2:0.1:0.1:0.2:0.2 | 0.3:0.1:0.1:0.3:0.1 | 1:1:1:1:1 | 0.3:0.2:0.3:0.1:0.2 | 0.3:0.3:0.3:0.3:0.5 | 0.3:0.3:0.3:0.3:0.5 | 0.3:0.3:0.5:0.3:0.3 | 0.3:0.3:0.2:0.3:0.5 | 0.5:C.3:0.5:0.2:0.3 |
| $C_4$ | 0.5:0.5:0.3:0.3:0.3 | 0.3:0.3:0.3:0.3:0.1 | 3:6:3:8:5 | 1:1:1:1:1 | 2:4:2:3:4 | 2:3:2:6:3 | 1:4:3:2:4 | 0.5:0.5:0.3:0.5:1 | 0.5:1:0.5:0.3:1 |
| $C_5$ | 0.5:0.2:0.3:0.3:0.5 | 0.3:0.2:0.3:0.2:0.1 | 3:3:3:4:2 | 0.5:0.3:0.5:0.3:0.3 | 1:1:1:1:1 | 1:1:2:2:1 | 1:2:4:3:2 | 0.5:0.3:0.3:1:0.3 | 0.5:1:0.5:0.3:1 |
| $C_6$ | 0.5:0.2:0.3:0.5:0.5 | 0.3:0.1:0.2:0.3:0.2 | 3:3:4:3:2 | 0.5:0.3:0.5:0.2:0.3 | 1:1:0.5:0.5:1 | 1:1:1:1:1 | 1:1:2:0.5:1 | 0.5:0.1:0.3:0.3:0.1 | 0.5:0.3:0.3:0.3:0.3 |
| $C_7$ | 1:0.3:0.3:0.5:0.3 | 0.5:0.3:0.2:0.1:0.1 | 3:4:2:3:3 | 1:0.3:0.3:0.5:0.3 | 1:0.5:0.3:0.3:0.5 | 1:1:0.5:2:1 | 1:1:1:1:1 | 0.5:0.3:0.5:0.3:0.2 | 0.5:0.2:0.5:0.3:0.3 |
| $C_8$ | 3:2:1:2 | 2:1:2:0.5:0.3 | 3:3:6:4:2 | 2:2:3:2:1 | 2:4:4:1:3 | 2:7:3:3:7 | 2:3:2:3:6 | 1:1:1:1 | 0.5:2:2:2:2 |
| $C_9$ | 1:0.5:0.3:0.5:0.5 | 1:0.3:1:1:0.3 | 2:4:2:5:3 | 2:1:2:3:1 | 2:1:2:3:1 | 2:4:3:3:3 | 2:5:2:3:3 | 2:0.5:0.5:0.5:0.5 | 1:1:1:1:1 |

**Table 2** $CR_k$ matrix of comparison and weights of experts

| Expert | $CR_k$ | $\delta_k$ | $w_k$ |
|--------|--------|------------|-------|
| E1 | 0.0695 | 14.3788 | 0.2346 |
| E2 | 0.0849 | 11.7718 | 0.1921 |
| E3 | 0.0980 | 10.2080 | 0.1666 |
| E4 | 0.0861 | 11.6210 | 0.1896 |
| E5 | 0.0752 | 13.2998 | 0.2170 |

*Step 3. Calculation of ARCM:*

ARCM is now estimated based on the data of Table 1 and using Eqs. (1)–(5), the components $\xi_{ij}^{(k)}$ of the CM $N^{(k)}$are transformed into rough number $RN(\xi_{ij}^{(k)}) = \left[\underline{Lim}(\xi_{ij}^{(k)}), \overline{Lim}(\xi_{ij}^{(k)})\right]$ and five rough matrices $N^{(k)}$ ($k = 1, 2, \dots 5$) are obtained. Determination of the rough elements of CM $N^{(1)}$, $N^{(2)}$, ..., $N^{(5)}$, for $C_1$–$C_3$ are illustrated here. For every $N^{(k)}$ matrix, rough sequences that make up the rough number $RN(\xi_{13}^{(k)}) = \left[\underline{Lim}(\xi_{13}^{(k)}), \overline{Lim}(\xi_{13}^{(k)})\right]$ are achieved. From Table 1, for $C_1$–$C_3$ position, the object class $\xi_{13}^{(k)}$ with five elements $\xi_{13}^{(k)} = \{6; 9; 8; 6; 5\}$ is selected. Now, using Eqs. (1)–(5), we determined the rough sequences as follows:

$$\underline{Lim}(6) = \frac{1}{3}(6+6+5) = 5.67, \overline{Lim}(6) = \frac{1}{4}(6+9+8+6) = 7.25;$$

$$\underline{Lim}(9) = \frac{1}{5}(6+9+8+6+5) = 6.80, \overline{Lim}(9) = 9;$$

$$\dots$$

$$\underline{Lim}(5) = 5, \overline{Lim}(5) = \frac{1}{5}(6+9+8+6+5) = 6.80;$$

This way, we get five rough sequences, as shown below:

$$RN(\xi_{13}^{(1)}) = [5.67, 7.25]; RN(\xi_{13}^{(2)}) = [6.8, 9.0]; RN(\xi_{13}^{(3)}) = [6.25, 8.5];$$
$$RN(\xi_{13}^{(4)}) = [5.67, 7.25]; RN(\xi_{13}^{(5)}) = [5.0, 6.8].$$

By applying Eq. (9) and the experts' weights of Table 2, we get the ARN as $RN(\xi_{13}) = \left[\underline{Lim}(\xi_{13}), \overline{Lim}(\xi_{13})\right] = \left[\xi_{13}^-, \xi_{13}^+\right]$

$$RN(\xi_{13}) = \left[\xi_{13}^-, \xi_{13}^+\right] = RN(\xi_{13}^{(1)}, \xi_{13}^{(2)}, \dots, \xi_{13}^{(5)})$$
$$= \begin{cases} \xi_{13}^- = (5.67)^{0.2346} \cdot (6.80)^{0.1921} \cdots (5.00)^{0.2170} \\ \xi_{ij}^+ = (7.25)^{0.2346} \cdot (9.0)^{0.1921} \cdots (6.80)^{0.2170} \end{cases}$$

This way, we obtained the values of Table 3.

*Step 4. Computation of the PV*

**Table 3** Averaged matrix

| Criteria | $C_1$ | $C_2$ | $C_3$ | $C_4$ | $C_5$ | $C_6$ | $C_7$ | $C_8$ | $C_9$ |
|---|---|---|---|---|---|---|---|---|---|
| $C_1$ | [1, 1] | [0.96, 4.61] | [5.81, 7.65] | [2.32, 2.82] | [2.38, 3.85] | [2.18, 3.32] | [1.76, 3.18] | [0.44, 0.67] | [1.57, 2.3] |
| $C_2$ | [0.37, 3.75] | [1, 1] | [5.1, 7.79] | [3.24, 4.76] | [3.91, 5.89] | [3.88, 5.9] | [3.46, 6.98] | [0.75, 1.97] | [1.38, 2.81] |
| $C_3$ | [0.13, 0.17] | [0.13, 0.21] | [1, 1] | [0.17, 0.29] | [0.31, 0.4] | [0.31, 0.4] | [0.3, 0.39] | [0.25, 0.39] | [0.28, 0.44] |
| $C_4$ | [0.36, 0.44] | [0.23, 0.32] | [3.69, 6.14] | [1, 1] | [2.43, 3.51] | [2.4, 3.95] | [1.85, 3.49] | [0.46, 0.68] | [0.5, 0.83] |
| $C_5$ | [0.28, 0.44] | [0.18, 0.28] | [2.61, 3.32] | [0.3, 0.43] | [1, 1] | [1.13, 1.59] | [1.63, 2.97] | [0.31, 0.64] | [0.5, 0.83] |
| $C_6$ | [0.33, 0.47] | [0.17, 0.27] | [2.6, 3.31] | [0.28, 0.44] | [0.68, 0.92] | [1, 1] | [0.82, 1.32] | [0.2, 0.37] | [0.31, 0.4] |
| $C_7$ | [0.34, 0.65] | [0.16, 0.33] | [2.65, 3.34] | [0.31, 0.64] | [0.37, 0.69] | [0.84, 1.34] | [1, 1] | [0.28, 0.44] | [0.3, 0.44] |
| $C_8$ | [1.62, 2.35] | [0.65, 1.6] | [2.69, 4.36] | [1.58, 2.3] | [1.85, 3.49] | [3.06, 5.57] | [2.41, 3.99] | [1, 1] | [1.28, 1.93] |
| $C_9$ | [0.46, 0.69] | [0.46, 0.86] | [2.4, 3.9] | [1.32, 2.2] | [1.32, 2.2] | [2.6, 3.32] | [2.37, 3.59] | [0.56, 0.99] | [1, 1] |

Based on the data of Table 3 and Eqs. (11) and (12), the normalized values of weights coefficients are computed, as given in Table 4.

The determination of the $C_1$–$C_3$ element of Table 4 is obtained by the following computations. Using Eq. (11), the following values are first calculated.

$$RN(\xi'_{13}) = \sum_{i=1}^{9} RN(\xi_{i3}) = \left[ \sum_{i=1}^{9} \xi_{i3}^-, \sum_{i=1}^{9} \xi_{i3}^+ \right]$$

$$= \begin{cases} \sum_{i=1}^{9} \xi_{i3}^- = 5.81 + 5.10 + \cdots + 2.40 = 28.56 \\ \sum_{i=1}^{9} \xi_{i3}^+ = 7.65 + 7.79 + \cdots + 3.90 = 40.81 \end{cases}$$

Then based on these values, we obtain rough number $RN(\xi'_{i3}) = \sum_{i=1}^{9} RN(\xi_{i3}) = [28.56, 40.81]$. These values are further used to normalize the third column of the average interval RCM. Thus, for the $C_1$–$C_3$ position, we obtain

$$RN(\varpi_{13}) = \frac{RN(\xi_{13})}{\sum_{i=1}^{9} RN(\xi_{i3})} = \frac{[5.81, 7.65]}{[28.56, 40.81]} = [0.14, 0.27]$$

Similarly, applying Eq. (11), the other values of Table 4 are achieved. Now using Eq. (14), the criteria weights are finally estimated. For example, the following values are calculated by dividing the weight of the first row of the normalized matrix by the number of criteria:

$$RN(w_1) = \left[ \frac{1}{9} \sum_{i=1}^{9} \varpi_{ij}^-, \frac{1}{9} \sum_{i=1}^{9} \varpi_{ij}^+ \right] = \frac{1}{9} \cdot [0.962, 3.055] = [0.107, 0.339]$$

The same process is followed for the other weight coefficients.

$RN(w_1) = [0.107, 0.339]$; $RN(w_2) = [0.130, 0.423]$; $RN(w_3) = [0.018, 0.042]$; $RN(w_4) = [0.066, 0.162]$; $RN(w_5) = [0.041, 0.104]$; $RN(w_6) = [0.032, 0.075]$; $RN(w_7) = [0.032, 0.083]$; $RN(w_8) = [0.106, 0.287]$; $RN(w_9) = [0.073, 0.184]$.

## 4.2  Implementation of R-CODAS Model

Now, the step-by-step implementation of the R-CODAS model is explained below.

*Step 1. Construct the basic RN decision matrix ($\Phi$)*

**Table 4** Normalized criteria weights

| Criteria | $C_1$ | $C_2$ | $C_3$ | $C_4$ | $C_5$ | $C_6$ | $C_7$ | $C_8$ | $C_9$ |
|---|---|---|---|---|---|---|---|---|---|
| $C_1$ | [0.10, 0.20] | [0.1, 0.17] | [0.14, 0.27] | [0.16, 0.27] | [0.11, 0.27] | [0.08, 0.19] | [0.07, 0.20] | [0.06, 0.16] | [0.14, 0.32] |
| $C_2$ | [0.04, 0.76] | [0.11, 0.25] | [0.12, 0.27] | [0.22, 0.45] | [0.18, 0.41] | [0.15, 0.34] | [0.13, 0.45] | [0.11, 0.46] | [0.13, 0.4] |
| $C_3$ | [0.01, 0.04] | [0.01, 0.05] | [0.02, 0.04] | [0.01, 0.03] | [0.01, 0.03] | [0.01, 0.02] | [0.01, 0.03] | [0.03, 0.09] | [0.03, 0.06] |
| $C_4$ | [0.04, 0.09] | [0.02, 0.08] | [0.09, 0.21] | [0.07, 0.09] | [0.11, 0.25] | [0.09, 0.23] | [0.07, 0.22] | [0.06, 0.16] | [0.05, 0.12] |
| $C_5$ | [0.03, 0.09] | [0.02, 0.07] | [0.06, 0.12] | [0.02, 0.04] | [0.05, 0.07] | [0.04, 0.09] | [0.06, 0.19] | [0.04, 0.15] | [0.05, 0.12] |
| $C_6$ | [0.03, 0.10] | [0.02, 0.07] | [0.06, 0.12] | [0.02, 0.04] | [0.03, 0.06] | [0.04, 0.06] | [0.03, 0.08] | [0.03, 0.09] | [0.03, 0.06] |
| $C_7$ | [0.03, 0.13] | [0.02, 0.08] | [0.06, 0.12] | [0.02, 0.06] | [0.02, 0.05] | [0.03, 0.08] | [0.04, 0.06] | [0.04, 0.1] | [0.03, 0.06] |
| $C_8$ | [0.16, 0.48] | [0.07, 0.41] | [0.07, 0.15] | [0.11, 0.22] | [0.08, 0.25] | [0.12, 0.32] | [0.09, 0.26] | [0.14, 0.24] | [0.12, 0.27] |
| $C_9$ | [0.05, 0.14] | [0.05, 0.22] | [0.06, 0.14] | [0.09, 0.21] | [0.06, 0.15] | [0.10, 0.19] | [0.09, 0.23] | [0.08, 0.23] | [0.09, 0.14] |

Results of the expert assessment for the considered biomaterials are shown in Table 5. By applying Eqs. (1)–(5), elements of the matrix are first translated into RNs, after which, by implementing Eq. (16), we average the RNs to obtain Table 6.

*Step 2. Normalization of basic matrix element* ($\Phi$)

The normalization of the criteria is carried out using Eq. (18). The first six criteria ($C_1$–$C_6$) belongs to the max (benefit) set, and three criteria ($C_7$–$C_9$) belongs to the min (cost) set. Normalized basic RN decision matrix is shown in Table 7.

*Step 3 and 4. Calculate RN-weighted normalized matrix and determine FR negative-ideal solution*

By the multiplication of Table 7 and the criteria weights, we obtain the following weighted matrix.

$$
R = \begin{bmatrix}
[0.02, 0.23] & [0.02, 0.23] & \cdots & [0.01, 0.07] \\
[0.02, 0.22] & [0.02, 0.26] & \cdots & [0.01, 0.07] \\
\vdots & \vdots & \ddots & \vdots \\
[0.01, 0.15] & [0.02, 0.25] & \cdots & [0.01, 0.05]
\end{bmatrix}_{11 \times 9}
$$

Now by applying Eq. (20), we obtain the RN negative-ideal solution matrix NS = $[\text{RN}(ns_j)]_{1 \times 9}$ as given below:

$$
\text{NS} = \begin{bmatrix}
\text{RN}(ns_1) = [0.011, 0.145]; & \text{RN}(ns_2) = [0.017, 0.234]; & \text{RN}(ns_3) = [0.000, 0.002]; \\
\text{RN}(ns_4) = [0.004, 0.026]; & \text{RN}(ns_5) = [0.002, 0.011]; & \text{RN}(ns_6) = [0.001, 0.008]; \\
\text{RN}(ns_7) = [0.001, 0.007]; & \text{RN}(ns_8) = [0.011, 0.087]; & \text{RN}(ns_9) = [0.005, 0.035];
\end{bmatrix}_{1 \times 9}
$$

*Step 5 and 6. Calculate the RN distances of alternatives from the RN negative-ideal solution and determine relative assessment matrix (RA)*

After calculation of the RN negative-ideal solution matrix NS = $[\text{RN}(ns_j)]_{1 \times 9}$, we obtain the $\text{ED}_i$ and $\text{HD}_i$ distances of the biomaterial alternatives from the RN negative-ideal solution, using Eqs. (21)–(24), as exhibited in Table 8.

In order to obtain the elements of the relative assessment matrix RA = $[p_{ik}]_{11 \times 11}$, we use previous obtained Euclidean distances and Hamming distances and Eqs. (26) and (27), respectively.

$$
\text{RA} = \begin{array}{c}
 \\
A_1 \\
A_2 \\
A_3 \\
\vdots \\
A_{11}
\end{array}
\begin{array}{cccccc}
A_1 & A_2 & A_3 & \cdots & A_{11} \\
\begin{bmatrix}
0.000 & -0.055 & -0.075 & \cdots & 0.090 \\
0.055 & 0.000 & -0.010 & \cdots & 0.145 \\
0.075 & 0.010 & 0.000 & \cdots & 0.165 \\
\vdots & \vdots & \vdots & \ddots & \vdots \\
-0.090 & -0.145 & -0.165 & \cdots & 0.000
\end{bmatrix}
\end{array}
$$

*Step 7. Calculate the assessment score ($AS_i$) of each biomaterial*

**Table 5** Experts' evaluation of biomaterials with respect to criteria

Expert 1

| Material | $C_1$ | $C_2$ | $C_3$ | $C_4$ | $C_5$ | $C_6$ | $C_7$ | $C_8$ | $C_9$ |
|---|---|---|---|---|---|---|---|---|---|
| Stainless steel 316 ($A_1$) | 10 | 7 | 517 | 350 | 8 | 8 | 200 | 8 | 1 |
| Stainless steel 317 ($A_2$) | 9 | 7 | 630 | 415 | 10 | 8.5 | 200 | 8 | 1.1 |
| Stainless steel 321 ($A_3$) | 9 | 7 | 610 | 410 | 10 | 8 | 200 | 7.9 | 1.1 |
| Stainless steel 347 ($A_4$) | 9 | 7 | 650 | 430 | 10 | 8.4 | 200 | 8 | 1.2 |
| Co–Cr alloy (castable) ($A_5$) | 10 | 9 | 655 | 425 | 2 | 10 | 238 | 8.3 | 3.7 |
| Co–Cr alloy (wrought) ($A_6$) | 10 | 9 | 896 | 600 | 10 | 10 | 242 | 9.1 | 4 |
| Pure titanium ($A_7$) | 8 | 10 | 550 | 315 | 7 | 8 | 110 | 4.5 | 1.7 |
| Ti–6Al–4V ($A_8$) | 8 | 10 | 985 | 490 | 7 | 8.3 | 124 | 4.4 | 1.9 |
| Epoxy-70% glass ($A_9$) | 7 | 7 | 680 | 200 | 3 | 7 | 22 | 2.1 | 3 |
| Epoxy-63% carbon ($A_{10}$) | 7 | 7 | 560 | 170 | 3 | 7.5 | 56 | 1.6 | 10 |
| Epoxy-62% aramid ($A_{11}$) | 7 | 7 | 430 | 130 | 3 | 7.5 | 29 | 1.4 | 5 |

…

Expert 5

| | $C_1$ | $C_2$ | $C_3$ | $C_4$ | $C_5$ | $C_6$ | $C_7$ | $C_8$ | $C_9$ |
|---|---|---|---|---|---|---|---|---|---|
| $A_1$ | 10 | 6 | 516 | 357 | 8 | 8.7 | 193 | 8.4 | 1.3 |
| $A_2$ | 10 | 7 | 633 | 412 | 10 | 9.3 | 201 | 8.3 | 1.1 |
| $A_3$ | 9 | 7 | 610 | 411 | 10 | 7.0 | 210 | 7.3 | 1.2 |

(continued)

**Table 5** (continued)

Expert 1

| Material | $C_1$ | $C_2$ | $C_3$ | $C_4$ | $C_5$ | $C_6$ | $C_7$ | $C_8$ | $C_9$ |
|----------|-------|-------|-------|-------|-------|-------|-------|-------|-------|
| $A_4$    | 9     | 7     | 562   | 425   | 10    | 7.7   | 209   | 8.6   | 1.3   |
| $A_5$    | 9     | 8     | 657   | 428   | 2     | 10.0  | 224   | 8.8   | 3.4   |
| $A_6$    | 9     | 8     | 897   | 598   | 10    | 7.9   | 247   | 9.6   | 3.8   |
| $A_7$    | 8     | 9     | 551   | 321   | 7     | 6.3   | 117   | 4.7   | 1.6   |
| $A_8$    | 8     | 9     | 987   | 483   | 7     | 4.5   | 118   | 4.4   | 1.7   |
| $A_9$    | 7     | 8     | 688   | 205   | 3     | 10.0  | 17    | 2.0   | 3.3   |
| $A_{10}$ | 6     | 8     | 566   | 169   | 3     | 7.0   | 53    | 1.7   | 9.7   |
| $A_{11}$ | 7     | 8     | 433   | 130   | 3     | 9.0   | 22    | 2.6   | 5.0   |

**Table 6** Averaged rough decision matrix

| Criteria/Alternative | $C_1$ | $C_2$ | $C_3$ | $C_4$ | $C_5$ | $C_6$ | $C_7$ | $C_8$ | $C_9$ |
|---|---|---|---|---|---|---|---|---|---|
| $A_1$ | [8.94, 9.83] | [6.75, 7.64] | [517.69, 522.11] | [347.46, 352.30] | [8.17, 8.69] | [8.17, 8.50] | [195.74, 199.13] | [7.78, 8.22] | [1.03, 1.18] |
| $A_2$ | [8.75, 9.64] | [7.17, 8.06] | [625.95, 638.45] | [412.49, 417.43] | [10.0, 10.0] | [8.70, 9.09] | [199.89, 200.66] | [8.02, 8.23] | [1.04, 1.12] |
| $A_3$ | [8.64, 8.96] | [7.47, 8.53] | [606.99, 612.36] | [405.85, 410.76] | [10.0, 10.2] | [6.65, 7.50] | [203.52, 208.56] | [7.60, 7.87] | [1.05, 1.15] |
| $A_4$ | [8.65, 9.35] | [7.36, 7.84] | [610.35, 649.43] | [425.22, 428.29] | [9.24, 9.81] | [7.85, 8.22] | [201.59, 206.09] | [8.09, 8.49] | [1.08, 1.21] |
| $A_5$ | [8.65, 9.35] | [8.47, 9.53] | [653.54, 658.62] | [422.98, 427.29] | [2.09, 2.37] | [9.47, 9.92] | [227.28, 234.32] | [8.33, 8.63] | [3.50, 3.66] |
| $A_6$ | [9.04, 9.36] | [8.36, 9.25] | [891.71, 895.08] | [595.21, 598.25] | [9.87, 9.97] | [8.43, 9.48] | [244.76, 248.78] | [9.28, 9.69] | [3.93, 4.11] |
| $A_7$ | [8.16, 8.64] | [8.36, 9.25] | [546.96, 551.21] | [312.85, 317.82] | [7.17, 7.68] | [6.74, 7.58] | [111.84, 115.40] | [4.01, 4.45] | [1.64, 1.72] |
| $A_8$ | [8.04, 8.36] | [9.16, 9.64] | [978.86, 985.28] | [485.79, 491.21] | [7.12, 7.48] | [5.42, 7.34] | [118.73, 121.77] | [4.28, 4.36] | [1.75, 1.85] |
| $A_9$ | [6.65, 7.35] | [6.94, 7.83] | [678.36, 688.02] | [198.07, 204.28] | [2.02, 2.76] | [7.94, 9.52] | [18.60, 21.14] | [1.99, 2.21] | [3.05, 3.20] |
| $A_{10}$ | [6.36, 7.25] | [7.36, 7.84] | [557.94, 566.44] | [165.38, 169.59] | [1.95, 2.74] | [6.42, 7.14] | [52.91, 55.11] | [1.49, 1.74] | [9.70, 9.90] |
| $A_{11}$ | [6.65, 7.35] | [7.36, 7.84] | [428.99, 432.29] | [127.77, 131.01] | [2.53, 2.88] | [7.90, 8.66] | [22.25, 25.90] | [1.70, 2.33] | [4.93, 5.11] |

**Table 7** Normalized rough decision matrix

| Criteria/Alternative | $C_1$ | $C_2$ | $C_3$ | $C_4$ | $C_5$ | $C_6$ | $C_7$ | $C_8$ | $C_9$ |
|---|---|---|---|---|---|---|---|---|---|
| $A_1$ | [0.19, 0.68] | [0.13, 0.55] | [0.02, 0.05] | [0.10, 0.24] | [0.07, 0.19] | [0.05, 0.13] | [0.04, 0.10] | [0.12, 0.35] | [0.14, 0.37] |
| $A_2$ | [0.18, 0.66] | [0.15, 0.61] | [0.02, 0.06] | [0.11, 0.26] | [0.08, 0.21] | [0.06, 0.14] | [0.04, 0.10] | [0.12, 0.35] | [0.15, 0.37] |
| $A_3$ | [0.18, 0.59] | [0.16, 0.68] | [0.02, 0.06] | [0.11, 0.26] | [0.08, 0.21] | [0.04, 0.11] | [0.04, 0.10] | [0.13, 0.36] | [0.15, 0.37] |
| $A_4$ | [0.18, 0.63] | [0.16, 0.58] | [0.02, 0.06] | [0.11, 0.26] | [0.08, 0.20] | [0.05, 0.12] | [0.04, 0.10] | [0.12, 0.34] | [0.14, 0.37] |
| $A_5$ | [0.18, 0.63] | [0.21, 0.83] | [0.02, 0.06] | [0.11, 0.26] | [0.04, 0.11] | [0.06, 0.15] | [0.03, 0.09] | [0.12, 0.34] | [0.12, 0.32] |
| $A_6$ | [0.19, 0.63] | [0.20, 0.79] | [0.03, 0.08] | [0.13, 0.32] | [0.08, 0.20] | [0.05, 0.14] | [0.03, 0.08] | [0.11, 0.30] | [0.12, 0.31] |
| $A_7$ | [0.16, 0.56] | [0.20, 0.79] | [0.02, 0.05] | [0.09, 0.23] | [0.07, 0.18] | [0.04, 0.11] | [0.05, 0.13] | [0.17, 0.49] | [0.14, 0.36] |
| $A_8$ | [0.16, 0.54] | [0.24, 0.85] | [0.04, 0.08] | [0.12, 0.29] | [0.07, 0.17] | [0.03, 0.11] | [0.05, 0.13] | [0.17, 0.48] | [0.14, 0.35] |
| $A_9$ | [0.12, 0.44] | [0.14, 0.58] | [0.03, 0.06] | [0.08, 0.19] | [0.04, 0.11] | [0.05, 0.14] | [0.06, 0.17] | [0.20, 0.56] | [0.13, 0.33] |
| $A_{10}$ | [0.11, 0.43] | [0.16, 0.58] | [0.02, 0.05] | [0.07, 0.18] | [0.04, 0.11] | [0.04, 0.10] | [0.06, 0.15] | [0.21, 0.57] | [0.07, 0.19] |
| $A_{11}$ | [0.12, 0.44] | [0.16, 0.58] | [0.02, 0.04] | [0.07, 0.16] | [0.04, 0.12] | [0.05, 0.13] | [0.06, 0.17] | [0.20, 0.57] | [0.11, 0.29] |

**Table 8** Euclidean and Hamming distances

Euclidean ($ED_i$) distances

| Criteria/Alternative | $C_1$ | $C_2$ | $C_3$ | $C_4$ | $C_5$ | $C_6$ | $C_7$ | $C_8$ | $C_9$ |
|---|---|---|---|---|---|---|---|---|---|
| $A_1$ | 0.0860 | 0.0000 | 0.0003 | 0.0125 | 0.0084 | 0.0016 | 0.0015 | 0.0153 | 0.0335 |
| $A_2$ | 0.0797 | 0.0261 | 0.0007 | 0.0161 | 0.0101 | 0.0023 | 0.0014 | 0.0129 | 0.0335 |
| $A_3$ | 0.0572 | 0.0554 | 0.0006 | 0.0157 | 0.0104 | 0.0003 | 0.0013 | 0.0171 | 0.0334 |
| $A_4$ | 0.0701 | 0.0129 | 0.0007 | 0.0167 | 0.0098 | 0.0012 | 0.0013 | 0.0121 | 0.0333 |
| $A_5$ | 0.0701 | 0.1175 | 0.0007 | 0.0167 | 0.0000 | 0.0033 | 0.0005 | 0.0097 | 0.0240 |
| $A_6$ | 0.0706 | 0.1000 | 0.0015 | 0.0263 | 0.0101 | 0.0027 | 0.0000 | 0.0000 | 0.0223 |
| $A_7$ | 0.0465 | 0.1000 | 0.0004 | 0.0105 | 0.0070 | 0.0004 | 0.0041 | 0.0536 | 0.0312 |
| $A_8$ | 0.0372 | 0.1244 | 0.0018 | 0.0203 | 0.0068 | 0.0000 | 0.0038 | 0.0509 | 0.0307 |
| $A_9$ | 0.0034 | 0.0118 | 0.0008 | 0.0041 | 0.0005 | 0.0028 | 0.0069 | 0.0742 | 0.0257 |
| $A_{10}$ | 0.0000 | 0.0129 | 0.0004 | 0.0022 | 0.0005 | 0.0003 | 0.0058 | 0.0792 | 0.0000 |
| $A_{11}$ | 0.0034 | 0.0129 | 0.0000 | 0.0000 | 0.0007 | 0.0017 | 0.0068 | 0.0771 | 0.0184 |

Hamming ($HD_i$) distances

| Criteria/Alternative | $C_1$ | $C_2$ | $C_3$ | $C_4$ | $C_5$ | $C_6$ | $C_7$ | $C_8$ | $C_9$ |
|---|---|---|---|---|---|---|---|---|---|
| $A_1$ | 0.0941 | 0.0000 | 0.0003 | 0.0143 | 0.0095 | 0.0021 | 0.0017 | 0.0171 | 0.0383 |
| $A_2$ | 0.0872 | 0.0284 | 0.0008 | 0.0186 | 0.0116 | 0.0029 | 0.0016 | 0.0147 | 0.0383 |
| $A_3$ | 0.0642 | 0.0594 | 0.0007 | 0.0181 | 0.0119 | 0.0005 | 0.0014 | 0.0194 | 0.0383 |

(continued)

**Table 8** (continued)

Euclidean (ED₁) distances

| Criteria/Alternative | $C_1$ | $C_2$ | $C_3$ | $C_4$ | $C_5$ | $C_6$ | $C_7$ | $C_8$ | $C_9$ |
|---|---|---|---|---|---|---|---|---|---|
| $A_4$ | 0.0772 | 0.0159 | 0.0008 | 0.0193 | 0.0112 | 0.0017 | 0.0015 | 0.0136 | 0.0381 |
| $A_5$ | 0.0772 | 0.1271 | 0.0009 | 0.0192 | 0.0000 | 0.0041 | 0.0006 | 0.0110 | 0.0274 |
| $A_6$ | 0.0788 | 0.1090 | 0.0017 | 0.0303 | 0.0115 | 0.0033 | 0.0000 | 0.0000 | 0.0255 |
| $A_7$ | 0.0520 | 0.1090 | 0.0004 | 0.0121 | 0.0080 | 0.0006 | 0.0046 | 0.0602 | 0.0357 |
| $A_8$ | 0.0423 | 0.1377 | 0.0021 | 0.0233 | 0.0077 | 0.0000 | 0.0044 | 0.0577 | 0.0352 |
| $A_9$ | 0.0042 | 0.0129 | 0.0010 | 0.0047 | 0.0005 | 0.0033 | 0.0078 | 0.0837 | 0.0294 |
| $A_{10}$ | 0.0000 | 0.0159 | 0.0005 | 0.0025 | 0.0005 | 0.0005 | 0.0066 | 0.0893 | 0.0000 |
| $A_{11}$ | 0.0042 | 0.0159 | 0.0000 | 0.0000 | 0.0008 | 0.0022 | 0.0077 | 0.0864 | 0.0211 |

**Table 9** Materials ranks using the RN-CODAS model

| Alternative | $H_i$ | Rank |
|---|---|---|
| $A_1$ | −0.650 | 8 |
| $A_2$ | −0.035 | 6 |
| $A_3$ | 0.167 | 5 |
| $A_4$ | −0.644 | 7 |
| $A_5$ | 1.376 | 3 |
| $A_6$ | 1.195 | 4 |
| $A_7$ | 1.644 | 2 |
| $A_8$ | 2.202 | 1 |
| $A_9$ | −1.408 | 9 |
| $A_{10}$ | −2.214 | 11 |
| $A_{11}$ | −1.634 | 10 |

RN-CODAS criteria function for the final ranking of the alternatives $A_i (i = 1, 2, \ldots, 11)$ are calculated using Eq. (27), as shown in Table 9.

## 5  Sensitivity Analysis, Discussion, and Validation

Sensitivity analysis (SA) is a key component for interpreting the outcomes of any multi-criteria analysis. The main objective of SA is to analyze the robustness of the proposed R-AHP-CODAS method. It aims to determine the minimum change I criteria weights that steers some changes in the ranking preorder of the biomaterial alternatives. Figure 2 shows the SA of the proposed R-AHP-CODAS model at varying weights of different material selection criteria for the considered case study. The evaluative outcome of the case study (Table 9) is analyzed through 36 different scenarios in which one criterion has been favored over others in each scenario by increasing its weight in the following manner. In the first scenario ($S_1$), criterion $C_1$ (tissue tolerance) was favored, in the second scenario ($S_2$), $C_2$ (corrosion resistance) was favoured, and so on. Variation in the ranking preorder of the biomaterials for different circumstances is revealed in Fig. 2.

From Fig. 2, it is well understood that the ranking preorder of the biomaterials has changed slightly due to the changes in criteria weights. From a comparison among the best two alternatives ($A_8$) in different scenarios with the initial rank of Table 9, it is noted that the best alternative ($A_8$) rank has not been affected by weight variations. Analysis of the ranking through 36 scenarios of Fig. 2 shows that alternative $A_8$ (Ti–6Al–4V) holds its rank in 32 scenarios (88.89%), while the second-best alternative (Pure titanium) holds its rank in 23 scenarios (63.89%). However, there are few minor changes in the ranking order for some intermediate biomaterials which is also confirmed by the standard deviation (SD) of the ranks, as shown in Fig. 3.

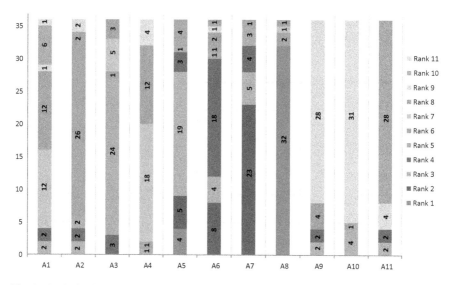

**Fig. 2** Analysis of biomaterials rankings through 36 scenarios

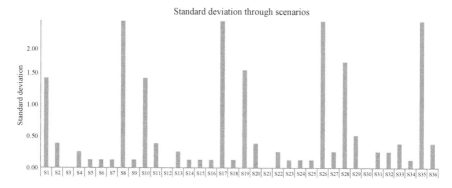

**Fig. 3** SD values for different scenarios

SD values are computed based on the initial ranks of Table 9 with that of achieved through different weight changing scenarios (Fig. 2). Figure 3 shows a high ranking correlation with a SD value of less than 0.50 in 28 scenarios. The mean value of SD for the scenarios for the biomaterial selection case study is found to be 0.428, which again signifies very good ranking agreement for all the scenarios.

To arrive at a final decision for selecting the optimal biomaterial, a comparative study has now been performed between different MCDM methods, namely COPRAS [26, 52] and MABAC [51, 53], as shown in Fig. 4. Ranking of the biomaterials according to these methods shows that alternative $A_8$ retained its first position for all the considered MCDM methods, thus establishing its superior acceptability over other biomaterial alternatives considered in the presented case study, as shown in.

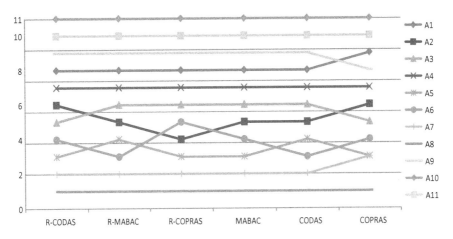

**Fig. 4** Comparison of ranks of biomaterials for different MCDM methods

**Table 10** Spearman's rank correlation coefficient values between R-AHP-CODAS and other MCDM methods

| MCDM method | Spearman's rank correlation coefficient |
|---|---|
| R-MABAC | 0.982 |
| R-COPRAS | 0.973 |
| MABAC | 0.991 |
| CODAS | 0.982 |
| COPRAS | 0.986 |
| Average SCC | 0.986 |

Ranking agreement and stability are validated by Spearman's rank correlation coefficient values, as exhibited in Table 10.

Table 10 shows a considerably high correlation between different MCDM methods. Spearman's rank correlation coefficient between the considered methods ranges from 0.982 to 1.00, which shows very strong correlation and ranking agreement among all these methods which ultimately establishes the reliability and credibility of the proposed model.

## 6   Conclusions

This paper proposes a new application of an integrated rough number-based AHP-CODAS model for selection of biomaterials. One real-life hip prosthesis joint material selection example demonstrates the potentiality and precision of the adopted model. The R-AHP method is used to determine the criteria weights, while alternative biomaterials are assessed by the R-CODAS model. In order to measure the quality of the results, weight SA and performance comparison with other well-established

MCDM methods have been carried out. Agreement between the obtained ranking orders is validated by using Spearman's rank correlation coefficients which indicates a very high rank correlation between all the considered methods, thus establishing the trustworthiness of the adopted approach. Amalgamation of fuzzy and neutrosophic theories with RNs can be the directions of future research.

# References

1. Mehrali M, Thakur A, Pennisi CP, Talebian S, Arpanaei A, Nikkhah M, Dolatshahi-Pirouz A (2017) Nanoreinforced hydrogels for tissue engineering: biomaterials that are compatible with load-bearing and electroactive tissues. Adv Mater 29(8):1–26
2. Huebsch N, Mooney DJ (2009) Inspiration and application in the evolution of biomaterials. Nature 462:426–432
3. Jahan A (2012) Material selection in biomedical applications: comparing the comprehensive VIKOR and goal programming models. Int J Mater Struct Integrity 6(2–4):230–240
4. Walker J, Shadanbaz S, Woodfield TB, Staiger MP, Dias GJ (2014) Magnesium biomaterials for orthopedic application: a review from a biological perspective. J Biomed Mater Res B Appl Biomater 102(6):1316–1331
5. Sridharan R, Cameron AR, Kelly DJ, Kearney CJ, O'Brien FJ (2015) Biomaterial based modulation of macrophage polarization: a review and suggested design principles. Mater Today 18(6):313–325
6. Chiti MC, Dolmans MM, Donnez J, Amorim CA (2017) Fibrin in reproductive tissue engineering: a review on its application as a biomaterial for fertility preservation. Ann Biomed Eng 45(7):1650–1663
7. Aherwar A, Singh AK, Patnaik A (2016) Current and future biocompatibility aspects of biomaterials for hip prosthesis. AIMS Bioeng 3(1):23–43
8. Bahraminasab M, Sahari BB (2013) NiTi shape memory alloys, promising materials in orthopedic applications. In: Shape memory alloys-processing, characterization and applications, InTech, USA, pp 261–278
9. Aherwar A, Singh A, Patnaik A, Unune D (2018). Selection of molybdenum-filled hip implant material using grey relational analysis method. In: Handbook of research on emergent applications of optimization algorithms, IGI Global, USA, pp 675–692
10. Hafezalkotob A, Hafezalkotob A (2017) Interval MULTIMOORA method with target values of attributes based on interval distance and preference degree: biomaterials selection. J Ind Eng Int 13(2):181–198
11. Farag MM (2013) Materials and process selection for engineering design. CRC Press, USA
12. Mousavi-Nasab SH, Sotoudeh-Anvari A (2017) A comprehensive MCDM-based approach using TOPSIS, COPRAS and DEA as an auxiliary tool for material selection problems. Mater Des 121:237–253
13. Panchal D, Singh AK, Chatterjee P, Zavadskas EK, Ghorabaee MK (2019) A new fuzzy methodology-based structured framework for RAM and risk analysis. Appl Soft Comput 74:242–254
14. Yazdani M, Chatterjee P, Zavadskas EK, Streimikiene D (2018) A novel integrated decision-making approach for evaluation and selection of renewable energy technologies. Clean Technol Environ Policy 20(2):403–420
15. Tavana M, Yazdani M, Di Caprio D (2017) An application of an integrated ANP–QFD framework for sustainable supplier selection. Int J Logistics Res Appl 20(3):254–275
16. Chatterjee P, Chakraborty S (2017) Development of a meta-model for determination of technological value of cotton fiber using design of experiments and TOPSIS method. J Nat Fibers 15(6)

17. Chatterjee P, Chakraborty S (2017) A developed meta-model for selection of cotton fabrics using design of experiments and TOPSIS method. J Inst Eng (India): Ser E 98(2):79–90
18. Chakraborty S, Chatterjee P, Prasad K (2018) An integrated DEMATEL-VIKOR method-based approach for cotton fibre selection and evaluation. J Inst Eng (India): Ser E 99(1):63–73
19. Chatterjee P, Chakraborty S (2016) A comparative analysis of VIKOR method and its variants. Decis Sci Lett 5(4):469–486
20. Ranjan R, Chatterjee P, Chakraborty S (2016) Performance evaluation of Indian railway zones using DEMATEL and VIKOR methods. Benchmarking: an Int J 23(1):78–95
21. Chatterjee P, Athawale VM, Chakraborty S (2010) Selection of industrial robots using compromise ranking and outranking methods. Robot Comput Integr Manuf 26(5):483–489
22. Kaklauskas A, Zavadskas EK, Raslanas S, Ginevicius R, Komka A, Malinauskas P (2006) Selection of low-e windows in retrofit of public buildings by applying multiple criteria method COPRAS: a Lithuanian case. Energy Build 38(5):454–462
23. Yazdani M, Chatterjee P, Zavadskas EK, Zolfani SH (2017) Integrated QFD-MCDM framework for green supplier selection. J Clean Prod 142(4):3728–3740
24. Maity SR, Chatterjee P, Chakraborty S (2012) Cutting tool material selection using grey complex proportional assessment method. Mater Des 36:372–378
25. Chatterjee P, Chakraborty S (2012) Materials selection using COPRAS and COPRAS-G methods. Int J Mater Struct Integrity 6(2/3/4):111–133
26. Chatterjee P, Athawale VM, Chakraborty S (2011) Materials selection using complex proportional assessment and evaluation of mixed data methods. Mater Des 32(2):851–860
27. Stanujkić D, Đorđević B, Đorđević M (2013) Comparative analysis of some prominent MCDM methods: a case of ranking Serbian banks. Serb J Manage 8(2):213–241
28. Ghorabaee MK, Zavadskas EK, Olfat L, Turskis Z (2015) Multi-criteria inventory classification using a new method of evaluation based on distance from average solution (EDAS). Informatica 26(3):435–451
29. Chatterjee P, Banerjee A, Mondal S, Boral S, Chakraborty S (2018) Development of a hybrid meta-model for materials selection using design of experiments and EDAS method. Eng Trans 66(2):187–207
30. Ghorabaee MK, Zavadskas EK, Turskis Z, Antucheviciene J (2016) A new combinative distance-based assessment (CODAS) method for multi-criteria decision-making. Econ Comput Econ Cybern Stud Res 50(3):25–44
31. Panchal D, Chatterjee P, Shukla RK, Choudhury T, Tamosaitiene J (2017) Integrated fuzzy AHP-CODAS framework for maintenance decision in urea fertilizer industry. Econ Comput Econ Cybern Stud Res 51(3):179–196
32. Bahraminasab M, Jahan A (2011) Material selection for femoral component of total knee replacement using comprehensive VIKOR. Mater Des 32(8):4471–4477
33. Jahan A, Edwards KL (2013) Weighting of dependent and target-based criteria for optimal decision-making in materials selection process: biomedical applications. Mater Des 49:1000–1008
34. Bahraminasab M, Sahari BB, Edwards KL, Farahmand F, Hong TS, Arumugam M, Jahan A (2014) Multi-objective design optimization of functionally graded material for the femoral component of a total knee replacement. Mater Des 53:159–173
35. Petković D, Madić M, Radenković G, Manić M, Trajanović M (2015) Decision support system for selection of the most suitable biomedical material. In: Proceedings of 5th international conference on information society and technology, Serbia, pp 27–31
36. Hafezalkotob A, Hafezalkotob A (2015) Comprehensive MULTIMOORA method with target based attributes and integrated significant coefficients for materials selection in biomedical applications. Mater Des 87:949–959
37. Chowdary Y, Sai Ram V, Nikhil EVS, Vamsi Krishna PNS, Nagaraju D (2016) Evaluation and prioritizing of biomaterials for the application of implantation in human body using fuzzy AHP AND TOPSIS. Int J Control Theory Appl 9(40):527–533
38. Kabir G, Lizu A (2016) Material selection for femoral component of total knee replacement integrating fuzzy AHP with PROMETHEE. J Intell Fuzzy Syst 30(6):3481–3493

39. Abd K, Hussein A, Ghafil A (2016) An intelligent approach for material selection of sensitive components based on fuzzy TOPSIS and sensitivity analysis. In: Proceedings of academics world 30th international conference, Australia, pp 13–18
40. Ristić M, Manić M, Mišić D, Kosanović M, Mitković M (2017) Implant material selection using expert system. Facta Univ, Ser: Mech Eng 15(1):133–144
41. Pamucar D, Lj Gigovic, Bajic Z, Janosevic M (2017) Location selection for wind farms using GIS multi-criteria hybrid model: an approach based on fuzzy and rough numbers. Sustainability 9(8):1–24
42. Song W, Ming X, Wu Z, Zhu B (2014) A rough TOPSIS approach for failure mode and effects analysis in uncertain environments. Qual Reliab Eng Int 30(4):473–486
43. Pamucar D, Mihajlovic M, Obradovic R, Atanaskovic P (2017) Novel approach to group multi-criteria decision making based on interval rough numbers: hybrid DEMATEL-ANP-MAIRCA model. Expert Syst Appl 88:58–80
44. Saaty TL, Vargas LG (2012) Models, methods, concepts and applications of the analytic hierarchy process. Springer Science and Business Media, pp 175
45. Saaty TL (1977) A scaling method for priorities in hierarchical structures. J Math Psychol 15(3):234–281
46. Saaty TL, Tran LT (2007) On the invalidity of fuzzifying numerical judgments in the analytic Hierarchy process. Math Comput Model 46(7):962–975
47. Stevic Z, Pamucar D, Vasiljevic M, Stojic G, Korica S (2017) Novel integrated multi-criteria model for supplier selection: case study construction company. Symmetry 9(11):1–34
48. Stevic Z, Pamucar D, Zavadskas EK, Cirovic G, Prentkovskis O (2017) The selection of wagons for the internal transport of a logistics company: a novel approach based on rough BWM and rough SAW methods. Symmetry 9(11):1–25
49. Ghorabaee MK, Amiri M, Zavadskas EK, Hooshmand R, Antuchevičienė J (2017) Fuzzy extension of the CODAS method for multi-criteria market segment evaluation. J Bus Econ Manage 18(1):1–19
50. Jahan A, Mustapha F, Md YI, Sapuan SM, Bahraminasab M (2011) A comprehensive VIKOR method for material selection. Mater Des 32(3):1215–1221
51. Pamucar D, Ćirović G (2015) The selection of transport and handling resources in logistics centers using Multi-Attributive Border Approximation Area Comparison (MABAC). Expert Syst Appl 42:3016–3028
52. Badi IA, Abdulshahed AM, Shetwan AG (2018) A case study of supplier selection for a steel-making company in Libya by using the Combinative Distance-based ASsessment (CODAS) model. Decis Mak: Appl Manage Eng 1(1):1–12
53. Mukhametzyanov I, Pamucar D (2018) A sensitivity analysis in MCDM problems: a statistical approach. Decis Mak: Appl Manage Eng 1(2):51–80

CPSIA information can be obtained
at www.ICGtesting.com
Printed in the USA
LVHW080210230919
631929LV00001B/2/P